The Open University

Technology Foundation Course Unit 6

The week number during which this unit should be studied is not necessarily the same as the unit number. Please consult your wall-chart study guide to the Technology Foundation Course to find the place of this unit in the course

MECHANICS
(MOVEMENT—ITS ORIGINS AND CONSEQUENCES)

Prepared by Keith Attenborough for the Technology Foundation Course Team

THE OPEN UNIVERSITY PRESS

The Technology Foundation Course Team

G. S. Holister (*Chairman and General Editor*)
K. Attenborough (*Engineering Mechanics*)
R. J. Beishon (*Systems*)
D. A. Blackburn (*Materials Science*)
J. K. Cannell (*Engineering Mechanics*)
A. Clow (*BBC*)
G. P. Copp (*Assistant Editor*)
D. G. Crabbe (*Course Assistant*)
C. L. Crickmay (*Design*)
N. G. Cross (*Design*)
E. S. L. Goldwyn (*BBC*)
J. G. Hargrave (*Electronics*)
R. D. Harrison (*Educational Technology*)
M. J. L. Hussey (*Engineering Mechanics*)
A. B. Jolly (*BBC*)
J. C. Jones (*Design*)
L. M. Jones (*Systems*)
R. D. R. Kyd (*Editor*)
J. McCloy (*BBC*)
R. McCormick (*Educational Technology*)
D. Nelson (*BBC*)
C. W. A. Newey (*Materials Science*)
S. Nicholson (*Design*)
G. Peters (*Systems*)
A. Porteous (*Engineering Mechanics*)
C. Robinson (*BBC*)
R. Roy (*Design*)
J. J. Sparkes (*Electronics*)
R. Thomas (*Economics*)
G. H. Weaver (*Materials Science*)
P. I. Zorkoczy (*Electronics*)
and the late Professor R. K. Ham (*Materials Science*)

Open University Press
Walton Hall Milton Keynes

First published 1972   Reprinted 1974 (with corrections)

Copyright © 1972 The Open University

All rights reserved. No part of this work may be reproduced in any form, by mimeograph or any other means, without permission in writing from the publishers.

Designed by the Media Development Group of the Open University

SBN 335 020504 8

Printed in Great Britain by
Martin Cadbury, a specialized division of Santype International
Worcester and London

Open University courses provide a method of study for independent learners through an integrated teaching system, including textual material, radio and television programmes and short residential courses. This text forms part of a series that makes up the correspondence element of the Foundation Course.

The Open University's courses represent a new system of University level education. Much of the teaching material is still in a developmental stage. Course and course materials are, therefore, kept continually under revision. It is intended to issue regular up-dating notes as and when the need arises, and new editions will be brought out when necessary.

For general availability of supporting material referred to in this book, please write to the Director of Marketing, The Open University, P.O. Box 81, Milton Keynes, MK7 6AT.

Further information on Open University courses may be obtained from The Admissions Office, The Open University, P.O. Box 48, Milton Keynes, MK7 6AB.

Contents

	What you have to do	4
	Objectives	5
1	**Introduction**	7
2	**Modelling in mechanics**	8
3	**Newton's laws of motion**	9
3.1	Pushes and pulls: force	9
3.2	Weight and mass	11
3.3	Another look at force: momentum	13
3.4	Action and reaction	18
3.4.1	Universal gravitation	19
3.4.2	Conservation of momentum	19
3.5	Motion in a curved path	21
4	**Work and energy**	25
4.1	Defining work	25
4.1.1	Force and displacement	25
4.1.2	Path independence	26
4.1.3	To work or not to work	26
4.1.4	Varying forces	26
4.2	Something for something	27
4.3	Storing energy	28
4.4	There's the rub	29
4.5	Simplifying the picture	30
4.6	Dissipation, conversion and conservation	30
4.7	The pile-driver	32
4.8	The advanced passenger train	33
4.9	The pendulum	34
5	**Simple harmonic motion**	35
5.1	Time for a swing	35
5.2	Damping	36
5.3	Strong damping and critical damping	36
6	**The kinetic theory of gases**	38
6.1	Introduction	38
6.2	Keeping up the pressure	38
6.3	From change in momentum to pressure	40
6.4	Meaning of temperature	42
6.5	A few complications	43
	Self-assessment questions	44
	Self-assessment answers and comments	48

Note to student

Where references are made in Course Units to the T100 Book of the Course *The Man-Made World*, we are using the following abbreviations:

T100/BK(Intro)	Introduction to *The Man-Made World*.
T100/BK1	Case Study 1. The electricity supply industry.
T100/BK2	Case Study 2. Education by satellite in Brazil.
T100/BK3	Case Study 3. The plastics and steel industries.
T100/BK4	Case Study 4. Transport.

What you have to do

The important thing with this unit is to make sure you understand the basic ideas which are described in non-mathematical language. We think it will help you to try a few *simple* calculations based on these ideas. These are given in 15 exercises distributed throughout the text, plus 15 self-assessment questions at the end. Try, if you can, to do the exercises without looking at the answers (cover them with a piece of paper), but don't spend too long on them – not more than 5 minutes on each – and don't worry if you can't do them.

Objectives

The intention of this unit is to introduce the principal terms and concepts used in mechanics so that they are available as tools elsewhere in the course. The unit assumes that Unit 0 (Modelling I) has provided you with the ability to follow simple mathematical arguments that involve mechanical ideas. It also assumes that you have gained a facility in translating verbal statements into mathematical statements and vice-versa.

You might have been able to do both of these operations before starting T100. However, I suggest that if you are meeting mathematical arguments with some trepidation you should keep Modelling I by your side—in particular that section devoted to manipulation (the Supplement).

After working through this unit you should be able to:

(1) Use correctly and explain the meaning of the following terms: force, mass, momentum, kinetic energy, potential energy, work, power, weight, dissipation, amplitude, periodic time, fluid pressure.

(2) State and explain in your own words the meaning of Newton's laws of motion, the conservation of momentum, and the conservation of energy.

(3) Calculate the work done during the movement of a simple force.

(4) Calculate the kinetic energy of a given body.

(5) Calculate the gravitational potential energy of a given body or fluid in a given configuration.

(6) Calculate
(a) the distance accumulated by a body after a certain time-interval when subject to a given force, by using Newton's laws.
(b) the speed gained by a body after a certain time-interval by using Newton's laws when given the force, or by using the conservation of energy when given the height above a reference level.

(7) Explain qualitatively the essential role of centripetal forces in rotational motion and quote the result $F = mv^2/r$.

(8) Calculate force as a rate of change of momentum and distinguish between a sudden impact and a cushioned impact by reference to the variation of force with time.

(9) Explain qualitatively the phenomenon of friction and calculate the work done in dragging an object across some surface given the relevant coefficient.

(10) Draw graphs showing the nature of damped oscillations when the damping is (a) small, (b) large, (c) critical. Give examples of simple harmonic motion and discuss both the importance of damping and choice of periodic time in mechanical systems.

(11) Explain the meaning of fluid pressure and temperature in terms of kinetic theory.

After you have completed the course, you should be able to illustrate your answers to questions based upon some of the above objectives with practical examples drawn from other units and the Case Studies.

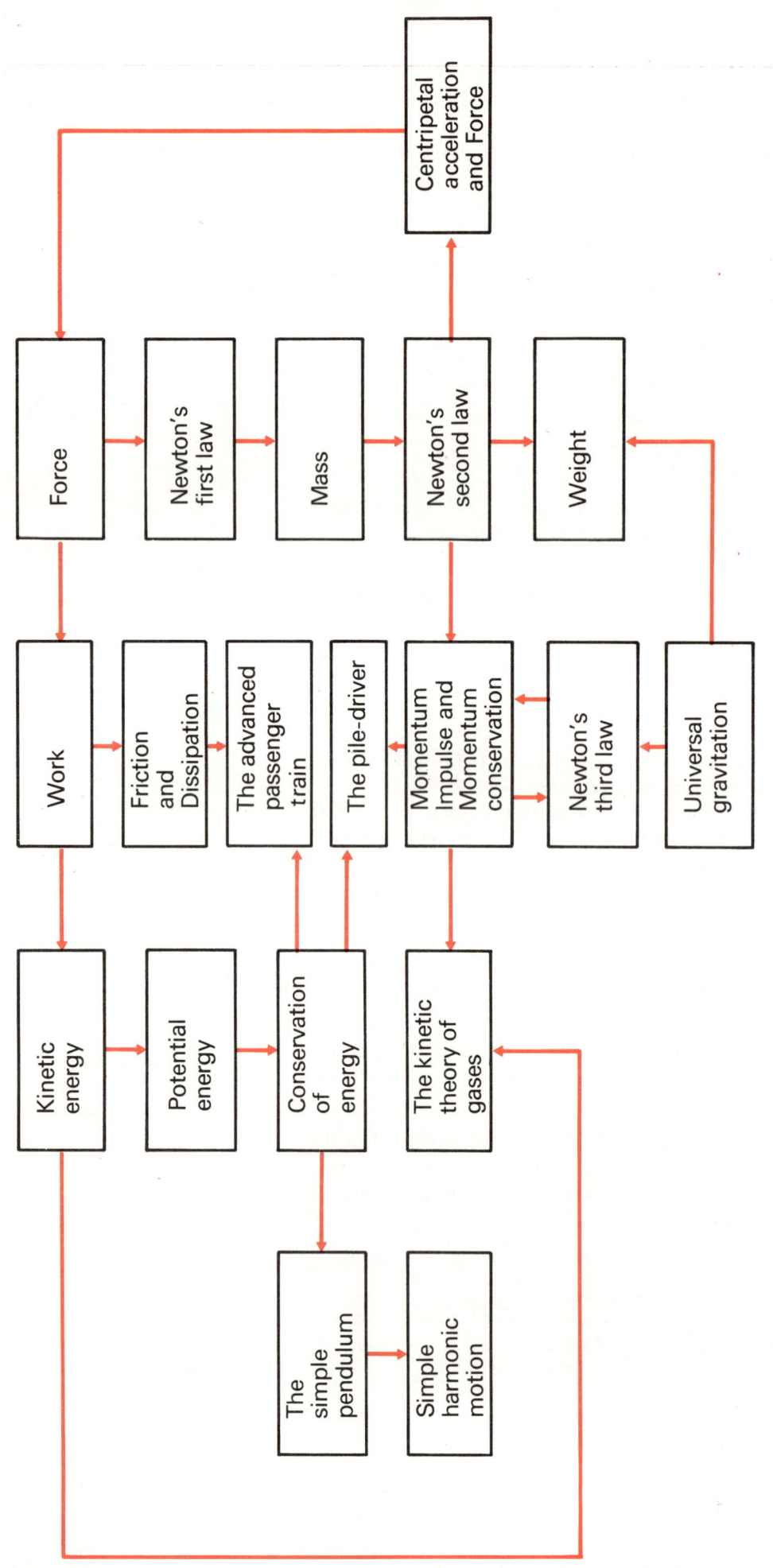

Section 1

Introduction

Much present technology is concerned with producing or preventing motion, from the acceleration of the smallest sub-nuclear particles (for example, electrons and protons) to the careful docking and manoeuvring of space capsules. We have to know how to describe and calculate the movements, in terms of the forces and energy involved.

In the Case Studies, you will have met particular examples. In putting satellites into orbit we must know the strengths, types and directions of the forces in order to ensure the satellite's stability. In generating power, we have to be acquainted with the various forms of energy and how to convert from one form to another. This problem is often linked with how to convert from one type of motion to another; for example, from motion in a straight line to motion in a circle, as when a water-wheel is driven by the flow of a river. The provision of transport necessarily involves mechanics—the mechanical design of advanced passenger trains or moving pavements, problems of isolation from the vibration caused by road or rail traffic, the scheduling of railway timetables and decisions on the locomotive power to use on particular runs. To make the choice between a steel and a plastic in a structure requires knowledge of the likely forces involved, and whether the particular structural member is in tension or in compression.

The study which provides us with this kind of information is called mechanics. Some knowledge of mechanics is essential for understanding much of what comes later in the course.

Figure 1 '*Much present technology is concerned with producing or preventing motion . . .*'

Section 2

Modelling in mechanics

In trying to model any mechanical system, considerable progress can be made by using graphical and mathematical methods. The basic elements of mechanical modelling and their elementary interactions are described in this unit. You will not meet many new elements or interactions even if you pursue a study of mechanics to a higher level.

The inferences that result from the formulation of a mechanical model can involve extensive mathematical calculations or numerical calculations which are only possible with the aid of computers. Typical problems include the determination of aerodynamic forces on aeroplanes in flight and the calculation of the movement of silt and sand caused by the currents and tides in river estuaries.

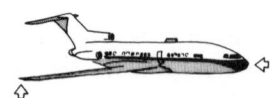

Complexity is not essential in a model. To an engineer a model is useful if it interprets observation and predicts behaviour.

Mathematics provides its own language for modelling and enables precise and unambiguous statements to be made, but we also need a non-mathematical vocabulary. Many of the words we use in mechanics are terms in everyday language, like force, work and energy. But when these words are used in mechanics they have a precision of meaning—an exactitude which is seldom encountered in ordinary conversation. For example, we have to confine ourselves to the sense of 'force' as 'that which tends to produce movement'. Nevertheless, while some of the ordinary intuitive associations are vague, and even occasionally misleading, others relate fairly closely to the sense in which they are used in mechanics and may help you towards a feeling for the subject. No one starts a study of mechanics from scratch. For example, you will already have a notion of 'heaviness' which is probably not far removed from the meaning of 'weight' in mechanics.

Section 3

Newton's laws of motion

3.1 Pushes and pulls: force

Some ideas of mechanics which you have met in Unit 0 (Modelling I) are those of displacement, velocity and acceleration. These concepts are related as shown in Figure 2 and with them you should be in a position to describe and predict any motion which occurs in a straight line.

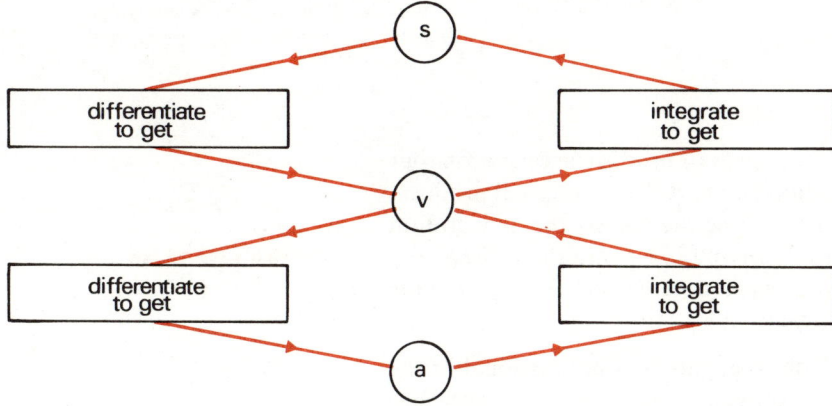

Figure 2 The use of integration and differentiation to relate distance, speed and acceleration.

Sir Isaac Newton first related force with change in motion. His first law may be stated as: *Each body continues in its state of rest or uniform motion in a straight line, unless compelled to change that state by forces which act upon it.*

Newton's first law

Uniform motion in a straight line is not observable. Everything that moves near the Earth's surface eventually comes to rest. If we look further than this, beyond the Earth's atmosphere, to an artificial satellite, to the Moon or to some other planet we cannot find anything moving—and continuing to move—in a straight line. This law is therefore a remarkable basis for the study of mechanics.

If you check back to the definitions of velocity and acceleration which you met in the modelling unit you will find that 'uniform motion in a straight line' corresponds to constant velocity, or, in other words, no acceleration. 'State of rest' also corresponds to a state of no acceleration. So the existence of acceleration inevitably means the presence of force. One could therefore define force as 'that which tends to produce acceleration'. The use of the verb 'tends' is meaningful.

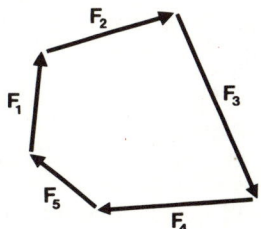

In Modelling I you met an example where a multiplicity of forces were acting and yet produced no motion. A post was buried in the ground and was pulled upon by a number of ropes. It was found that when the resulting displacement of the top of the post was zero, the *resultant* of the forces was zero as well. In other words, the law refers really to *resultant* force; with the undeflected post there is no resultant. When a number of forces are involved, motion will result only if the forces are unbalanced, and will be in the direction of the resultant of the forces. Any single force will *tend* to produce acceleration unless it is balanced by an equal and opposite force. Only an *unbalanced* force produces acceleration.

resultant

unbalanced force

If these forces are not producing acceleration, what are they producing?

A state of stress in the ropes.

Exercise 1

If the rate of change (with respect to time) of a particle's displacement is constant, is a force acting upon the particle?

When a particle is moving with uniform motion in a straight line, then its displacement is altering at a constant rate; however, we know from Newton first law that in this case there is no force acting since there is no acceleration. An unbalanced force exists when the velocity is changing.

So Newton's first law tells us that force tends to produce acceleration, but it does not tell us how much acceleration any particular force will produce. Intuitively we know that the greater the force the greater the acceleration that is likely to be produced. Newton's second law states that: *Change of motion is proportional to the force applied and is made in the direction of the straight line in which that force is applied.*

Newton's second law

The idea that the change of motion or acceleration is proportional to the force means that if you double the force, you produce twice the acceleration; if you halve the force, then only half the acceleration will result—and so on. Furthermore, the same force applied to a heavy body will produce less acceleration than when it is applied to a light body.

We call the measure of an object's reluctance to move the *mass* of the object. The terms *inertia* or inertial mass are often used as alternatives. The use of the word inertia in this context is similar to its use in everyday language. Likewise, when we think of something as 'massive' we imagine it to be hard to move—we have this idea of inertia at the back of our minds.

mass
inertia

If a body is made throughout of a uniform substance, its mass is proportional to its volume, so size is an indicator of mass in such cases. Newton, in fact, referred to mass as 'quantity of matter' which is a more reliable expression when a mixture of substances is involved.

The dual dependence of force on both mass and acceleration can be expressed

$$\text{force} = (\text{mass}) \; times \; (\text{acceleration})$$

or, symbolically,

$$\boldsymbol{F} = m\boldsymbol{a} \dots \dots \dots \dots \dots \dots (1)$$

The bold type is used to remind you that since acceleration has direction as well as magnitude (i.e. it is a vector), force also must be a vector with the same direction as the acceleration it produces. This is the usual form for Newton's second law of motion. You will have noticed that, in Modelling I, acceleration was measured in units of (m/s)/s. A shorthand way of writing this is m s^{-2}. The unit of mass in the system of units that is used in this course, is the *kilogramme* (kg). (The standard Kilogramme itself is a lump of platinum which is kept at Sèvres in France.) In order to define a unit of force so that the constant or proportionality may be eliminated from Newton's second law, we choose that force which will produce unit acceleration when applied to unit mass. This unit is called

kilogramme

the *newton* (N). A force of 1 newton (1 N) causes an acceleration of 1 m s^{-2} in a mass of 1 kg.*

newton

Exercise 2

What is the constant force required to decelerate an oil tanker of mass 100 000 000 kg from 7·2 m s^{-1} to rest in 1 hour?

The tanker has to slow down from 7·2 m s^{-1} in 3 600 s. This means that it must slow down by $\frac{7\cdot2}{3\,600}$ m s^{-1} each second. The size of the acceleration† is therefore 0·002 m s^{-2}.
The force that must be exerted on the tanker is the product of the mass of the tanker and the acceleration. That is
100 000 000 × 0·002 = 200 000 N

3.2 Weight and mass

The most common force that we experience is that due to *gravitational attraction*. All bodies fall with the same acceleration. This is a fact that Galileo is reputed to have demonstrated by dropping objects of different weights from the leaning tower of Pisa. So the force due to gravitational attraction is unusual in that it always produces the same acceleration of any body on which it acts. What is the implication of Newton's second law in this case? If the constant acceleration due to gravity is represented by g, then the force (F) on mass (m) is $F = mg$ and so the force is proportional to the mass. Of course, we don't use the phrase, 'force due to gravitational attraction on a body' very frequently in either everyday language or mechanics. We refer to the *weight* of a body.

weight

Exercise 3

(1) What will be the speed after 3 seconds of a stone which is released from rest and allowed to fall freely under gravity?

(2) How far will it have fallen?

(The acceleration due to gravity near the Earth's surface is 9·81 m s^{-2}.)

(1) 29·4 m s^{-1}
(2) 44·1 m

The acceleration is 9·8 (m/s)/s approximately. The speed therefore is increasing by 9·8 m/s every second. Since the stone starts from rest, the speed is therefore 9·8 m/s after 1 second, 2×9·8 = 19·6 m/s after 2 seconds and 3×9·8 = 29·4 m/s after 3 seconds. The distance accumulated is equal to the area underneath the speed/time curve (which is a straight line) and therefore you need to find the area of the triangle shown in Figure 3. This is ½ × 3 × 29·4 = 44·1 m.

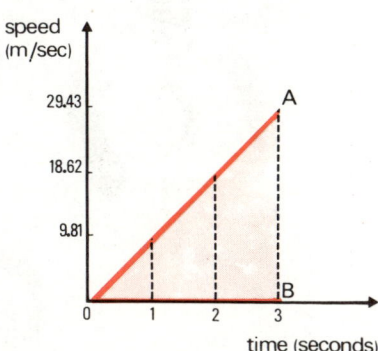

Figure 3 Speed against time for a body falling freely under gravity (for three seconds).

* Many other systems of units of mass and force are in use. One with which you might be familiar is the pound-foot-second system. The multiplicity of systems of units is often a nuisance and sometimes responsible for numerical mistakes. In this unit we shall adhere to a single system to avoid any risk of confusion. There is a tendency towards the universal adoption of this system of units (SI units).

† In mechanics we use the word acceleration to refer to both 'acceleration' and 'deceleration' as they are used in everyday language. If we are using vectors then a negative sign indicates deceleration.

The fact that weight is proportional to mass enables us to compare masses by comparing weights. This is what we do when we use the (scale) balance or gravity balance.

Alternatively, we can use the fact that a length of metal or a spring extends by an amount proportional to the magnitude of the weight suspended from it. (This is a consequence of Hooke's law which you met in Modelling I.) So the extension of the metal can be calibrated in terms of force. The size of the force due to gravity can be written as the mass multiplied by the acceleration due to gravity. Once we know the force all we have to do is to divide by the acceleration due to gravity (g) to obtain the mass. This is the method we use when we use the *spring balance*.

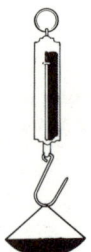

Exercise 4

What is the weight of 1 kg (mass)?

Since the acceleration due to gravity is 9·81 m s^{-2} and the weight of an object is the force on that object due to gravitational attraction, then the weight of 1 kg must be $1 \times 9.81 = 9.81$ N.

If the spring balance is calibrated in newtons, it is handy for measuring unknown forces. The form of the spring balance will depend upon the size of the force. The type of balance used for weighing luggage or potatoes is not of much use in much engineering where very large forces are involved. A different type of spring called a *proving ring* is often used (Figure 4).

proving ring

Figure 4 Proving rings. The dial gauge in the centre of each ring measures the force applied across the ring.

To sum up—as a unit of mass we have settled on the kilogramme. Mass is a measure of the reluctance of a body to respond to force. As the unit of force, we use the newton. One newton produces an acceleration of 1 m s^{-2} when it acts on a mass of 1 kg. With the aid of g, the acceleration due to gravity, we can base a method of measuring the mass of a body upon the measurement of its weight.

summary

3.3 Another look at force : momentum

Suppose that a car (mass 750 kg) travelling at 50 km/h were to drive into the rear of a stationary goods vehicle (mass 7 500 kg). Suppose further that the collision were to take place on a level road, and the handbrake of the goods vehicle was not applied. What would happen?

In this type of collision the motion of both vehicles is altered, and the vehicles are deformed, so forces must be acting on both the car and the goods vehicle, but it would be difficult to ascertain these forces at any time during the collision. Let us look at the effect of the collision on the goods vehicle. We know the mass of the vehicle is 7 500 kg. Let us assume that the goods vehicle moves off after the collision with a speed of 5 km/h.

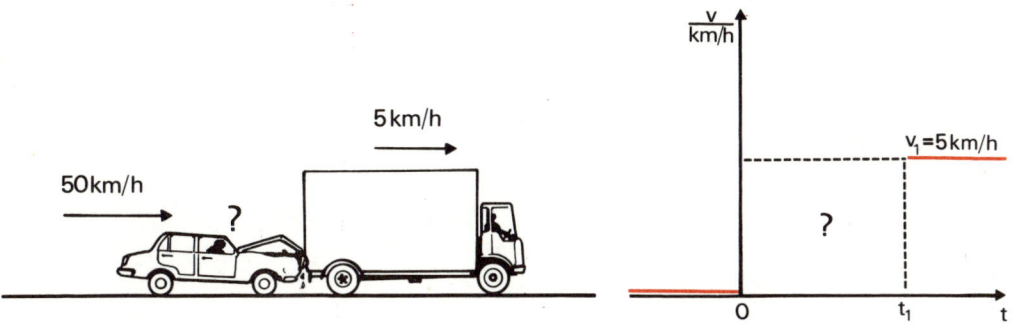

Figure 5 (a) *Collision between a saloon car and a stationary goods vehicle.* (b) *The known values of speed against time for this collision.*

This is represented diagrammatically in Figure 5(a) and in graphical form as in Figure 5(b). I have taken the time t as equal to zero just before the impact, when the velocity of the goods vehicle is also zero. At some time t_1 after the collision we can see the goods vehicle moving away at 5 km/h (we will concern ourselves with the subsequent behaviour of the car later). We don't necessarily know anything about the forces on the goods vehicle or its speed between the times 0 and t_1 but perhaps the simplest assumption we can make is that the velocity increases as a straight line between these two points (Figure 6). In this case, we are actually in a position to determine the force from Newton's second law.

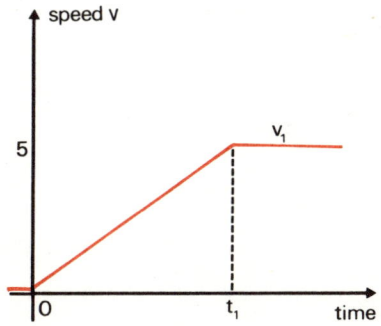

Figure 6 A simple assumed form for the way speed varies with time during the collision.

Force is mass times acceleration. You should remember from Modelling I that the acceleration is the slope of the speed/time curve. This will be given by the difference between the two values of speed at $t = 0$ and $t = t_1$ divided by the time-interval t_1. If we represent the mass of the goods vehicle by m, then the equation for the force is

$$F = \frac{mv_1}{t_1} \dots\dots\dots\dots\dots(2)$$

We can write this in a different way by multiplying both sides of the equation by t_1.

$$Ft_1 = mv_1 \dots\dots\dots\dots\dots\dots(3)$$

You will notice that the right-hand side of this equation contains only mass and velocity. We call the product of mass and velocity, *momentum*, so mv_1 is the change in momentum of the goods vehicle. The left-hand side of the equation represents the product of the force and the time during which it acts. This product is called the *impulse* of the force.

momentum

impulse

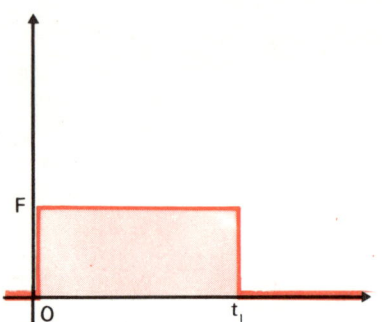 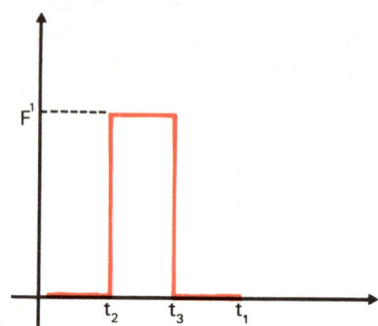

Figure 7 The variation of force with time corresponding to the speed/time relationship shown in Figure 6.

Figure 8 Another way in which the force could vary with time in the collision between saloon car and goods vehicle.

As we have assumed the value of mv_1 we can calculate the product of the force and the time over which it acts, irrespective of the shape we assume for the speed/time graph in between 0 and t_1. Since in Figure 6 we have assumed the slope of the speed/time graph to be constant between $t = 0$ and t_1, the acceleration must be constant between these times. As force is mass times acceleration, the force must also be constant over this time interval, and the change in momentum mv_1 ($= Ft_1$) must therefore be equal to the area of the rectangle shaded in Figure 7. In other words, the change in momentum is equal to the area underneath the force/time graph. The same area could result if the force were much greater but were to act for a much shorter time, as in Figure 8. If in the latter case the force were to be F' and the time interval were to be that between t_2 and t_3 (the different suffixes on the t indicate different times) then we would have

$$Ft_1 = F'(t_3 - t_2) = mv_1 \dots\dots\dots\dots(4)$$

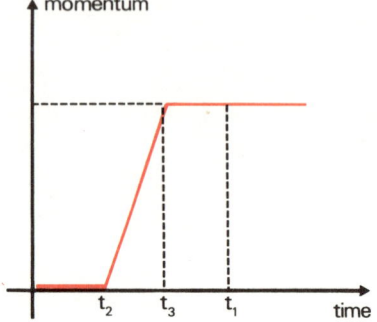

Figure 9 The momentum change corresponding to the impulse in Figure 8.

14

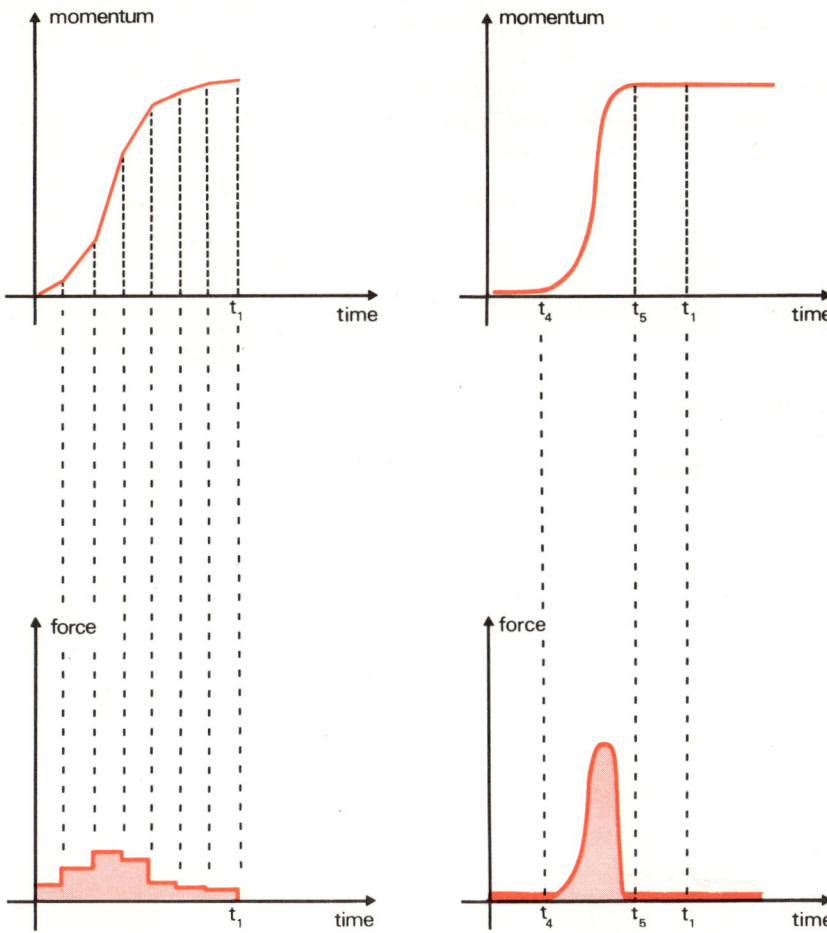

Figure 10 *A more complicated variation in momentum corresponding to the collision.*

Figure 11 *The corresponding variation of force with time.*

Figure 12 *A variation of momentum with time which has a form more appropriate to a collision.*

Figure 13 *The corresponding force/time curve.*

In this case what will the momentum/time graph look like? Since the force is larger, the acceleration is larger, and so the slope of the velocity/time graph is larger. Momentum is velocity multiplied by mass, so this will mean a steep slope in a momentum/time graph (Figure 9). We have assumed, so far, that the velocity of the goods vehicle increases uniformly with time. A less restrictive assumption would be that the velocity changes uniformly during successive short time-intervals, giving rise to a momentum/time graph like that of Figure 10. The corresponding 'staircase' force/time graph is shown in Figure 11. The total area underneath this curve would again be equal to $m\mathbf{v}_1$. If we then progress to a momentum/time graph with a slope that varies continually between t_4 and t_5 (Figure 12) and a corresponding force/time graph (Figure 13) we reach the most likely curves for the collision. The force/time curve must again encompass the same area ($= m\mathbf{v}_1$) if it is to result in the assumed velocity (5 km/h) of the goods vehicle after impact.

> What expression would you write down for the area underneath the curve in Figure 13 in terms of force and time?

You should remember from Modelling I that you need to write down the integral of force with respect to time between the times t_4 and t_5. This is the most general expression for impulse.

This means that 'the integral of force with respect to time is equal to the change in momentum' and points to a new expression for force. The force/time graph, in fact, results from a plot of the slope of the momentum/time curve. So *force is equal to the rate of change of momentum*. This is another form of Newton's second law, and is evident from the fact that force = mass × acceleration, which may be written $F = \dfrac{m\mathrm{d}v}{\mathrm{d}t} = \dfrac{\mathrm{d}}{\mathrm{d}t}(mv)$ (when m is constant) and which is the rate of change of momentum. (You should remember from Modelling I that $\dfrac{\mathrm{d}}{\mathrm{d}t}$ means differentiation with respect to time.)

When momentum change takes place very rapidly then the forces involved are very great and can cause considerable damage. For this reason, in a road accident of the type we have been considering, the occupants of the car are less liable to suffer severe injury if their car is designed so that progressive collapse of the front end occurs on impact. This ensures a more gradual change of momentum and reduced decelerating forces on the occupants. For the same reasons they are less likely to be injured if they are wearing seat belts—the restraining forces of the belts are less severe than the forces involved in hitting the windshield.

Exercise 5

Draw speed/time and force/time diagrams to show the forms of impulse involved in:
(a) a straight drive with a cricket bat;
(b) braking a car to a halt;
(c) accelerating a lift from rest.

Hints: don't worry too much about the absolute values of speed and force involved. Just try to think about the directions involved with these velocities and the likely size of the forces on, in turn, the *cricket ball*, the *car* and the *lift*.

Answer

(a) If we pick up the cricket ball's movement just before the bat hits it then it will be moving, say, with some velocity v_1 to the right. The impact will be short and sharp, in other words the impulse will take place over a very short time interval. After the impulse, the ball will have reversed its direction and will be moving to the left. We can represent speeds to the left as negative, speeds to the right as positive, so the speed after impact may be designated $-v_2$, as illustrated in Figure 14 and 15.

(b) Let us assume that the car is moving at speed v_3 before we start braking. We know that after braking the car is at rest, speed zero. Clearly it is best to break the car gently under normal circumstances, so the period of time between the start of braking and the finish of braking will be much longer than that involved during the impulse on the cricket bat. A typical result is shown in Figures 16 and 17.

(c) Most lifts are designed to give their occupants the least possible disturbance, so, as with the braking car, the impulse will be well spread out. However, since the lift is accelerating, the speed/time diagram would effectively be the reverse of that for the car. This is shown in Figures 18 and 19. In Figure 19 the force that is plotted is only the force needed to change the motion of the lift. There is an additional constant force in the lift-cables that counters the weight of the lift and its contents.

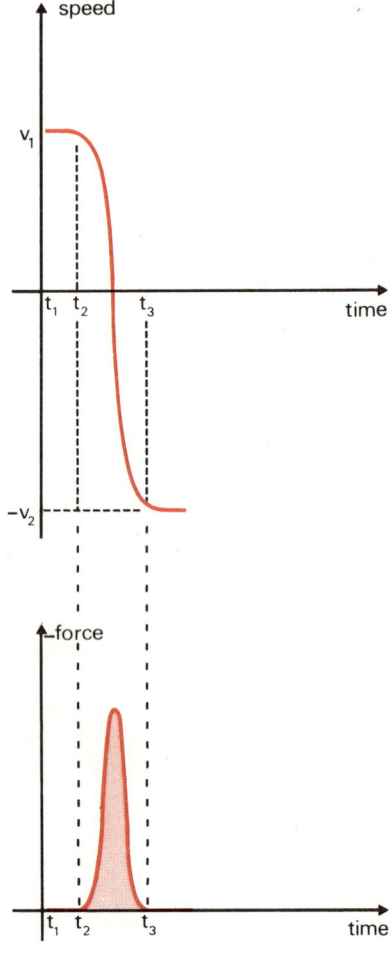

Figure 14 The speed/time curve of a cricket ball which is driven straight.

Figure 15 The corresponding force/time curve. Note that the sign of v does not concern us—only its rate of change.

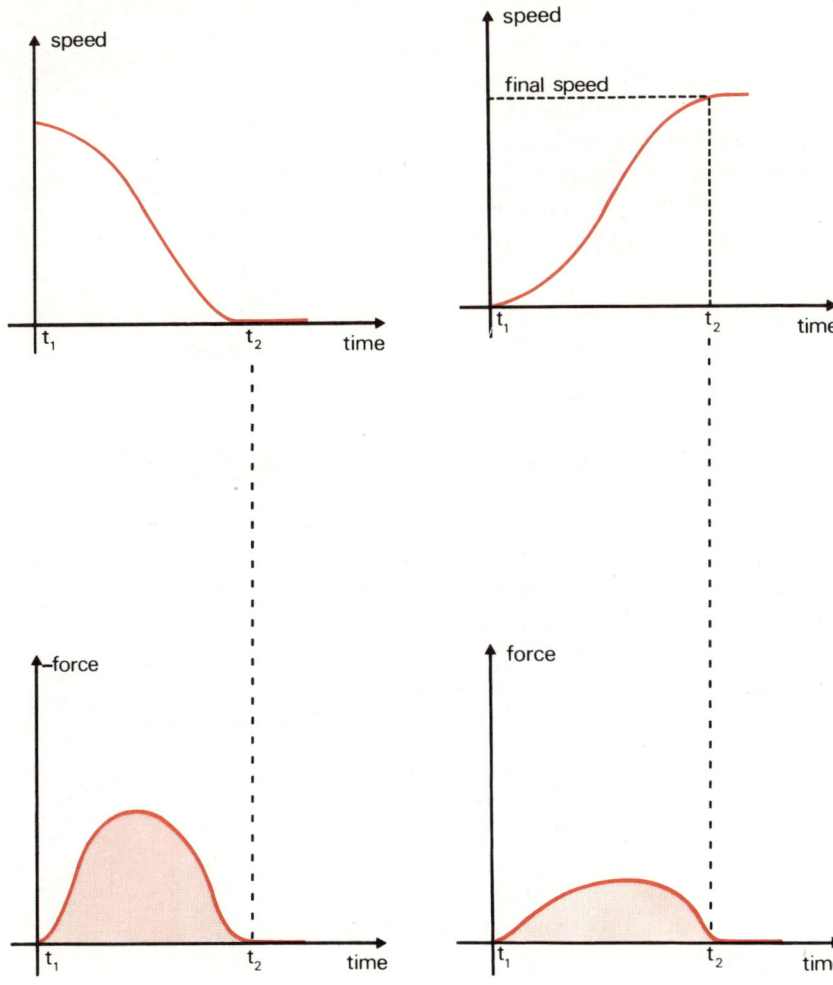

Figure 16 *The speed/time curve appropriate to a car which is braked to rest.*

Figure 17 *The corresponding force/time curve.*

Figure 18 *The speed/time curve of an accelerating lift.*

Figure 19 *The corresponding force/time curve. Note how the different velocity and time curves in Figures* 16 *and* 18 *produce almost identical force/time curves.*

Exercise 6

What is the constant force required to slow down an oil tanker of total mass 100 000 000 kg from 32 km/h to rest in 1 h?

Answer

The total change in momentum required is from the initial value, which is the mass of the tanker multiplied by its speed (m s^{-1}), to the final value of zero. So the total momentum change required is $\dfrac{100\,000\,000 \times 32 \times 1\,000}{3\,600}$

The factor 1 000 converts from kilometres to metres and the factor 3 600 converts from hours to seconds. This momentum change must be effected within 1 h so the rate of change of momentum, i.e. the change in momentum per second, must be the value we have already worked out divided by 3 600. Since the rate of change of momentum is the force required, the final value is approximately 250 000 N.

When considering the force required (over a certain time-interval) to bring an object to rest, the useful measure, as you will have seen from the above examples, is the initial momentum of the object. The fact that momentum is the product of mass and velocity means that a certain momentum can be the result of small mass and large velocity (as in a bullet fired from a rifle) or of large mass and relatively small velocity (as in a tanker at sea). The greater the momentum of a body the greater the force required to stop the body within a certain time-interval. Modern oil tankers have to start slowing up some 16 km before stopping. The mechanical idea of momentum is not clearly reflected in everyday language, where we use the word most often in contexts better suited to the mechanical idea of kinetic energy, which will be described later in this unit.

3.4 Action and reaction

So far we have looked mainly at the effects of forces on the motion of bodies in accordance with the first and second laws of motion. We have said very little about the origins of forces. One obvious source is in the contact of objects with each other. Here we find that exertion on our part can result in more than one force. If, for example, you push on a table, you experience a resistance. When a cricketer puts bat to ball, his wrists are jarred because the ball applies a force to the bat just as much as the bat applies a force to the ball. Newton suggested that the pairing of forces that occurs in these examples applies in all circumstances, whatever the kinds of forces that are involved. He took this as his third law of motion: *Action and reaction are equal and opposite.* **Newton's third law**

When the main rotor of a helicopter is in motion in a clockwise direction the engine must be driving it, that is exerting a force, in order to make the rotor move. This constitutes an *action*. There must consequently be a *reaction* on the engine mounting and the main fuselage, causing them to rotate in an *anti*clockwise direction. The *tail rotor*, which rotates about a horizontal axis in a vertical plane, counteracts this tendency, and is therefore a necessary part of the helicopter design (Figure 20). **reaction**

Of course, the driving of the tail rotor itself constitutes an action. How do you think the corresponding reaction tends to act? Why is this action not noticeable?

The essential assertion here is that all forces occur in pairs.

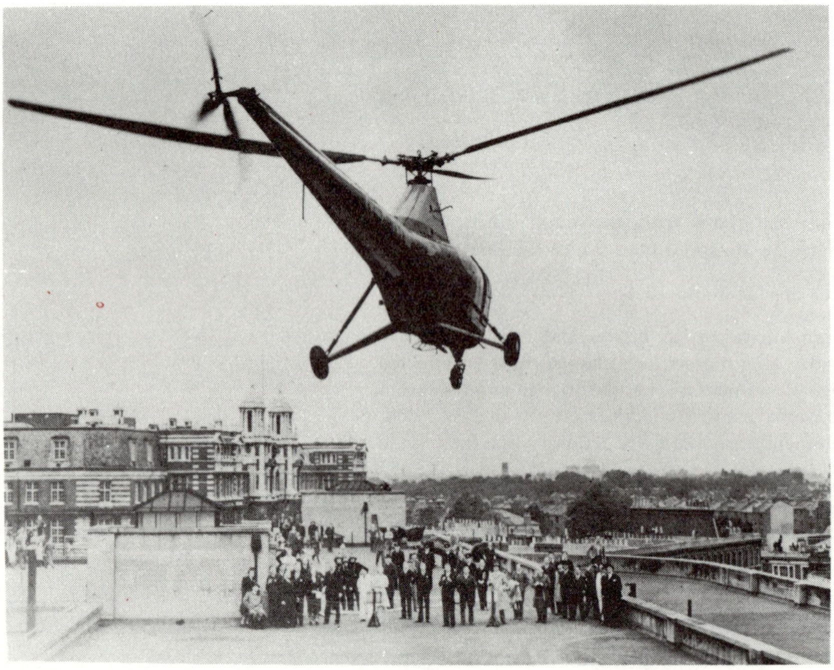

Figure 20   *A helicopter in flight. Note the tail rotor.*

3.4.1 Universal gravitation

I remarked earlier that the Earth exerts a gravitational attraction on all masses in its vicinity. Newton suggested that all masses exert a gravitational attraction on each other. The exact form of Newton's postulation for the magnitude of the gravitational force F between two masses m_1 and m_2 was that the force of attraction is equal to some constant multiplied by the product of the masses m_1 and m_2 and divided by the square of the distance between the masses. Symbolically this can be expressed as

$$F = \frac{Gm_1m_2}{r^2} \quad \ldots \ldots \ldots \ldots \ldots \ldots (5)$$

where r is the distance between the masses and in the unit system that we are using and the gravitational constant G has a value of $6 \cdot 67 \times 10^{-11}$ m^3 kg^{-1} s^{-2}. If the Earth is one of the masses, say m_1, we would expect the force on m_2 to be the weight, gm_2. We can do a rough check on this. When the Earth is one of the masses we measure the distance of any other mass from the centre of the Earth, so r in equation (5) must be the radius of the Earth, which is approximately 6 400 km. The mass of the Earth is approximately 6×10^{24} kg. So the value of $\frac{Gm_1}{r^2}$ is $6 \cdot 67 \times 10^{-11} \times 6 \times 10^{24}$ all divided by $6\,400 \times 6\,400 \times 10^6$. The 10^6 in the denominator occurs because we have to convert from km^2 to m^2. The result is the known value of g which is $9 \cdot 81$ m s^{-2}.

In choosing the Earth's radius as the effective distance between the mass of the Earth and that of an object on the Earth's surface, we assumed that, for the purpose of calculating the attraction due to gravity, the entire mass of the Earth can be regarded as concentrated at its centre. The justification for this need not detain us.

3.4.2 Conservation of momentum

Take two 2p pieces. Place them some distance apart on a table with a smooth surface. Flick one coin gently straight at the centre of the other. Unless you flick too hard, or not exactly at the centre of the second coin, one will come to rest and the other will move on some distance. Try the same action again, this time flicking a 10p piece against a $\frac{1}{2}$p. Usually, the 10p will carry on a little way after impact, the $\frac{1}{2}$p being knocked forward with considerable velocity.

The results of these simple experiments are illustrated in Figures 21 and 22. Now irrespective of which of these experiments we consider, if we call the coin which is fired A and the target B then we can call the force B exerts on A at any instant $_BF_A$. If the force is constant throughout the time of the collision, the total impulse on A is simply the product of this force and the time for which it acts. If the force is not constant throughout the time of the collision—and we cannot guarantee that it is—we must write the impulse (I_A) as the integral of the force with respect to time over the period for which the force acts (T)

$$I_A = \int_0^T {_BF_A} \, dt \ldots \ldots \ldots \ldots \ldots \ldots (6)$$

You should remember from Section 3.3 that the impulse on coin A should be equal to its change in momentum. The change of momentum is the product of the mass of the coin A and the change in its velocity. Similarly, the impulses experienced by coin B will depend upon the force that coin A exerts on coin B at any time. We can write this force as $_AF_B$. The time over which this force acts is the same as T because it is the time during which the coins are in contact. The integral of this force with respect to time over the relevant time-interval must equal the change in momentum of coin B.

From Newton's third law, at every instant between $t = 0$ and $t = T$, $_BF_A$ and $_AF_B$ should be equal in magnitude and opposite in direction. If this is the case, then, for constant forces,

$$_BF_A T = {_AF_B} T \dots \dots \dots \dots \dots (7)$$

and for forces which vary

$$\int_0^T {_BF_A}\, dt = \int_0^T {_AF_B}\, dt \dots \dots \dots \dots (8)$$

In other words the change in momentum of coin B and the change in momentum of coin A are equal. Since coin A slows down on impact and coin B speeds up, we can say that the momentum lost by coin A equals the momentum gained by coin B. *The total momentum is conserved.*

conservation of momentum

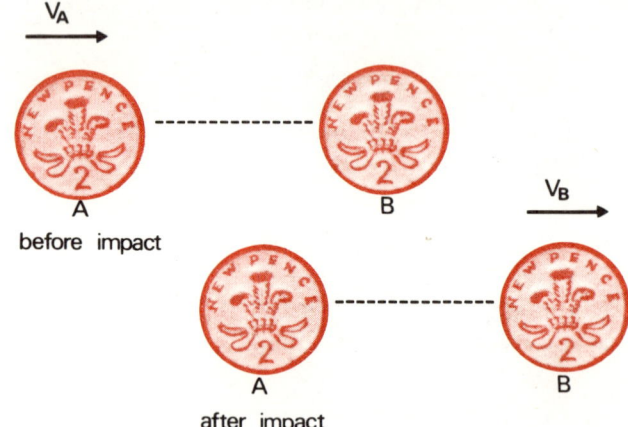

Figure 21 The collision of a 2p piece (A) moving with velocity v_A, with a stationary 2p piece (B).

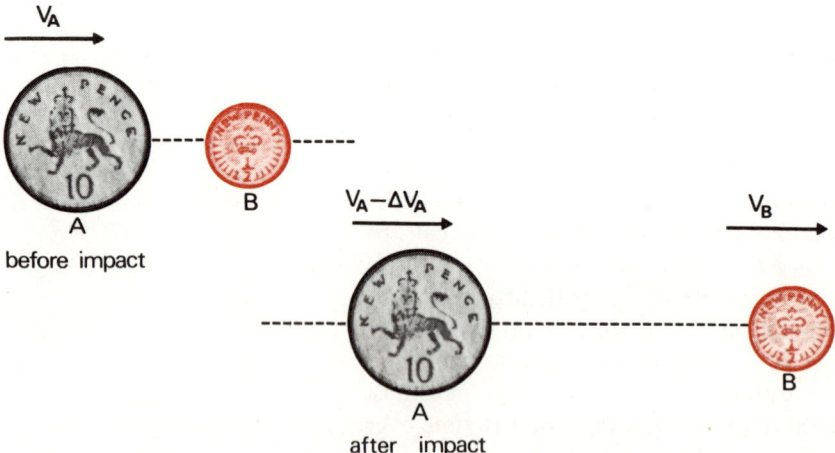

Figure 22 The collision of a 10p piece (A) moving with velocity v_A, with a $\frac{1}{2}$p (B).

This is a very useful idea as it enables us to calculate the speeds involved in an impact without knowing anything about the forces involved. For example, if in the second experiment the 10p piece has ten times the mass of the $\frac{1}{2}$p piece, and the 10p piece is moving at a speed of 3 m s^{-1} before impact, and 2·4 m s^{-1} after impact, we can calculate the velocity of the $\frac{1}{2}$p piece after impact.

Conservation of total momentum means that the total momentum before impact must equal the total momentum afterwards. Before the 10p coin strikes the ½p coin the total momentum is that of the 10p coin. If we represent the mass of the ½p coin by m, then the momentum of the 10p coin before impact is $3 \times 10 \times m = 30m$. After impact both the 10p coin and the ½p coin are moving and we have to add their individual momenta. After impact the speed of the 10p coin is $2\cdot 4$ m s^{-1} and its new momentum is $2\cdot 4 \times 10 \times m$, which is $24m$. The momentum of the ½p coin is the unknown speed $(v) \times m$. So by applying the principle of conservation of momentum we can write down the equation $30m = 24m + vm$.

Since m occurs on both sides of this equation, we can remove it altogether, giving us $v = 6$. The speed of the ½p coin after impact is 6 m s^{-1}. We were able to do this calculation without knowing the exact values of the masses of the coins. Knowledge of their ratio was sufficient.

The fact that momentum is the product of mass and velocity and that velocity has direction as well as magnitude means that momentum itself must have both direction and magnitude. We can represent the momentum vectors by lines whose length corresponds to the magnitude of the momentum, and whose direction corresponds to the direction of momentum. All the momenta in the cases so far described have been in the same direction. However, if the momenta are in different directions, then they will add according to the triangle rule, as do displacements, velocities and forces (see Modelling I).

Let us summarize the main ideas of Section 3 so far, in the order in which they have been presented:

summary—force and momentum

(1) The acceleration of an object is produced by an unbalanced force.

(2) The magnitude of the force accelerating an object is the product of the object's mass and acceleration. The force is in the direction of the acceleration.

(3) If an object experiences a force, then the force is equal to the rate of change of the object's momentum.

(4) Forces occur in equal and opposite pairs.

(5) When two objects collide, then their total momentum before collision is equal to their total momentum after collision.

Exercise 7

Turn back to Figure 5(a). Note that we put a question-mark over the car after impact, indicating we were not sure of its subsequent motion. You should now be in a position to calculate this. Do so now.

The car is brought to rest.

3.5 Motion in a curved path

Imagine swinging a heavy weight round and round on the end of a piece of cable, as an athlete does in preparing to throw the hammer. You know that if you let go the weight will immediately stop moving round you and will fly off at a tangent to the circle—that, after all, is the whole principle of hammer-throwing.

Before the weight is released it is moving, let us say, in a circle round your head and it must be the pull that you exert on it through the cable that keeps it moving in this way. When you let go you stop applying this force and the weight stops moving in a circle. The circular motion is thus a consequence of the inward pull of the cable on the weight.

Now according to Newton's second law of motion a force produces an acceleration of the body on which it acts. Because we have just seen that to keep a body moving in a circle there must always be a force pulling the body towards the centre of the circle we can deduce that the body is continually accelerating towards the centre of the circle. The acceleration of a body towards the centre of a circular path is called the centripetal (centre-seeking) acceleration, and the force producing centripetal acceleration is called the centripetal force.

It may, at first, seem a bit odd that a body moving steadily round and round a circle has a sustained acceleration towards the centre—certainly it never reaches the centre, nor does its speed change. But acceleration is the rate of change of velocity, and a velocity is a speed in a particular *direction*. A change in the *direction* of a body's motion thus implies a change in velocity. This in turn implies an acceleration even though the body's *speed* is unaltered.

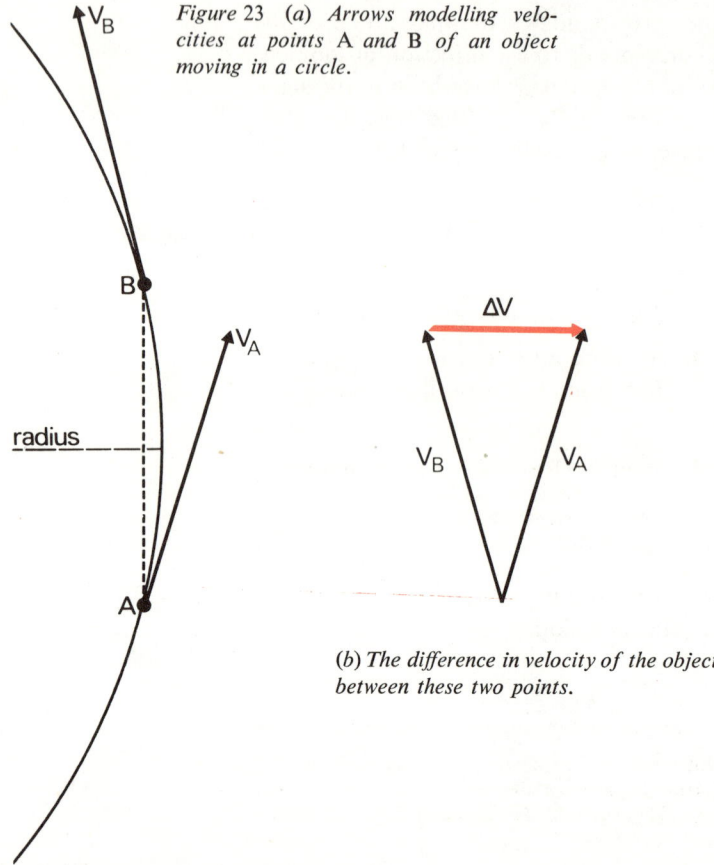

Figure 23 (*a*) *Arrows modelling velocities at points* A *and* B *of an object moving in a circle.*

(*b*) *The difference in velocity of the object between these two points.*

Figure 23 shows this graphically. It shows the positions, A and B, that a body moving in a circle occupies at the beginning and end of a short time-interval. We can model the velocities at A and B by drawing arrows of the appropriate size and direction at these two points. The velocity at A is represented by v_A and the velocity at B is represented by v_B. Although the *speed* of the object is unaltered between A and B, the *direction of movement* has changed, i.e. the velocity of the object has changed. To find the change in velocity, we have to discover what velocity must be added to the velocity at A to produce the velocity at B. How would you evaluate this change in velocity from what you remember about the way in which velocities add? You should remember that the triangle law must be applied. This has been done in Figure 23(b), where Δv denotes the change in velocity, that is the velocity which must be added to that at A to produce the velocity at B. You will see that I have chosen two points A and B, symmetrically disposed on either side of the radius, as shown in Figure 23(a). The change in velocity is directed along this radius. I could have chosen a much smaller time-interval so that the points A and B were extremely close together, one on either side of the point where this radius cuts the circle. The change in velocity in such a small time interval would, of course, be smaller, but would still be *directed* along the radius. The average acceleration for the short interval—found by dividing the (directed) change in velocity during the interval by the duration of the interval—is, therefore, also directed along the radius, and for shorter and shorter intervals, as A and B tend to coincide at a single point, this average acceleration becomes the centripetal acceleration whose existence we inferred from the mechanics of hammer-throwing.

> *An aside on centrifugal force.* We have just seen that a centripetal force is needed to make a mass move along a curved path. The question 'When is a path curved?' seems at first to present no problems but, in fact, the answer depends on your point of view. If you were to take a meal in the revolving restaurant on the GPO Tower you would be inclined to say that London was revolving around you. You would regard the tables and chairs in the restaurant as remaining at rest. We would then say that your *frame of reference* was rotating with the restaurant. Rotating frames of reference can be used when dealing with problems of rotating machinery like turbines and flywheels. You can see at once that Newton's laws of motion are not applicable in a rotating frame of reference because observations made from the revolving restaurant show Nelson's Column, for example, to be moving in a closed orbit. The motions of bodies as they are observed from a rotating frame of reference can, however, be predicted from a law of motion that is very like Newton's second law. This more complicated law can be expressed in terms of certain special forces, including *centrifugal* (centre-fleeing) forces of which you may have heard.
>
> We do not need to consider rotating frames of reference in this course and therefore have no use for the concept of centrifugal force. It is important, though, that a clear distinction should be made between centripetal and centrifugal forces.
>
> You may have realized, in reading the preceding remarks, that because the Earth itself is rotating, Newton's laws of motion are not applicable to motions observed from frames of reference on the Earth. Fortunately the Earth's rotation is so slow that for most purposes we can disregard it without upsetting the preductive value of our dynamical calculations. An interesting exception is in meteorology, where full account must be taken of the Earth's rotation to explain why, instead of blowing straight from the outer region of high barometric pressure to the inner region of low barometric pressure, the wind tends to circulate *round* a depression.

A few very simple experiments with rotating bodies tell us that the centripetal force increases with both the mass of the body and the speed of rotation but decreases with an increase of the object's distance from the centre round which it is moving. In fact, the centripetal force can be shown to equal the product of the mass of the object with the square of its speed, divided by the distance to the centre of its path. Symbolically, where the force is represented by F, the mass by m, the speed by v and the radius of the circular path by r,

$$F = \frac{mv^2}{r} \dots\dots\dots\dots\dots\dots(9)$$

With this information we are in a position to tackle a problem introduced in the case study on education by satellite in Brazil (T100/BK2). The problem is: what is the radius of a *circular orbit* round which the satellite will move once in 24 hours? In this case the necessary force, the centripetal force, is provided by the Earth's gravitational attraction.

orbit

If we represent the height of the satellite above the Earth's surface (which we want to calculate) by h (metres), then the distance of the satellite from the centre of the Earth (radius R) is $(R+h)$. So the centripetal force is given by the expression.

$$\frac{mv^2}{R+h}$$

where m represents the mass of the satellite. We can now remind ourselves of Newton's formula for gravitational attraction between two masses m_1 and m_2 which is

$$\frac{Gm_1m_2}{r^2}$$

If we write the mass of the earth as m_E and remember that the distance r also represents the distance from the centre of the Earth to the satellite, the equivalence of the centripetal force and that due to gravitational attraction gives us an equation:

$$\frac{mv^2}{(R+h)} = \frac{Gmm_E}{(R+h)^2} \quad \ldots \ldots \ldots \ldots (10)$$

where m_E represents the mass of the Earth (in kg). Since m appears on both sides of the equation, it can be cancelled. The calculation then becomes independent of the mass of the satellite, as pointed out in the case study.

The satellite must orbit once every 24 hours, that is, once every $24 \times 3\,600$ seconds. This means that the satellite must cover the distance $2\pi(R+h)$—which is the circumference of its orbit—in this time. In other words the satellite's speed must be $\dfrac{2\pi(R+h)}{(24 \times 3\,600)}$. We can substitute this in equation (10) to give us

$$\frac{4\pi^2(R+h)^2}{(R+h) \times 24^2 \times (3\,600)^2} = \frac{Gm_E}{(R+h)^2} \quad \ldots \ldots (11)$$

The $(R+h)$ in the denominator of the left-hand side can be cancelled with an $(R+h)$ from the numerator. Furthermore, cross-multiplying both sides of the equation by both denominators gives

$$4\pi^2(R+h)^3 = Gm_E \times 24^2 \times (3\,600)^2$$

Substitution of the relevant figures (quoted in Section 3.4.1) together with the value 10 for π^2 (this is approximate, but good enough in this case) enables us to write

$$(R+h)^3 = \frac{6 \cdot 67 \times 10^{-11} \times 6 \times 10^{24} \times 24^2 \times (3\,600)^2}{40} \quad \ldots \ldots (12)$$

This in turn gives a value of $1 \cdot 174 \times 36 \times 10^6$ m for $R+h$, which indicates that h is approximately 36 000 000 m—as quoted in the case study (T100/BK2).

Section 4

Work and energy

4.1 Defining work

4.1.1 Force and displacement

If a man loads a lorry with bricks, it is not sufficient for him merely to exert the force on each brick necessary to counteract its weight. He must maintain this force on each brick while he raises it through the necessary height, which is that of the tailboard on the lorry. The weight of each brick acts vertically downwards, so the force necessary to counteract this weight must act vertically upwards. When the man raises the bricks, therefore, he is moving them in the same direction as the force he exerts (Figure 24).

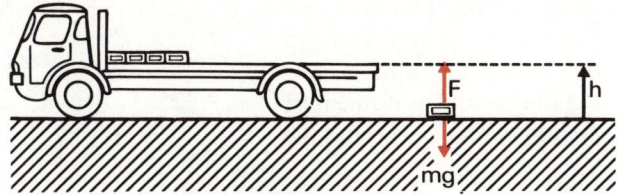

Figure 24 Forces on a brick which is about to be raised on to the tailboard of a lorry.

It is obvious that the greater the number of bricks, the heavier the bricks or the higher the tailboard of the lorry, the harder the man has to work. This means that any measure of *work* we adopt should simultaneously depend upon both the force exerted and the distance moved in the direction of the force.

work

In mechanics, *work* (W) *is defined as the product of the force* (F) *and the displacement* (s) *in the direction of the force.* Symbolically, we can write this $W = Fs$.

What is the work done on a brick of mass m in moving it through a height h?

The weight of the brick is mg. The force needed to counteract this weight must have magnitude mg and act in the opposite direction (vertically upwards). The vertical displacement is h, so the work done is mgh.

Exercise 8

If the mass of a brick is 4 kg, what is the work done in raising 40 bricks through 1·5 m?

Since the mass of the brick is 4 kg, the weight is $4 \times 9\cdot81$ N. The latter quantity, therefore, is the magnitude of the force needed to counteract the weight.
The work done in this case is the force multiplied by the height for each brick, so for 40 bricks
$W = 40 \times 4 \times 9\cdot81 \times 1\cdot5 = 2\,340$ N m.

Note the units for work in this example. Since it is the product of a force and a distance, the units are newton metres (N m) which are usually called *joules* (J).

joules

4.1.2 Path independence

When any brick is lifted, the path it takes before being placed on the lorry does not influence the measure of the work done in putting it there (*mgh*). For example, if the pile of bricks were some distance from the lorry, the man could first load the bricks on to a platform. He could then haul the platform to the required height *h* by means of a rope and pulley (which can slide on a wire) and push the platform—maintaining it at height *h*—along to the lorry. If the pulley slides smoothly along the wire no extra work is done by the man in this stage. The total work done is still *mgh* multiplied by the number of bricks. The general conclusion from this is that *work done against gravity is path independent*.

4.1.3 To work or not to work

It is important to realize that not everything we call 'work' in everyday conversation satisfies the mechanical definition. If we push on a wall or hold a book in front of us with arms outstretched for long enough our arms get tired. In both cases, you would probably say that we have been doing work. Indeed, as far as your body is concerned, with the chemical energy it is consuming, you might well have done 'work'. In mechanics we are interested usually in work that causes movement of masses from one place to another. So whether you have done *mechanical work* or not is determined by this interest. If we are interested only in the movement of the wall or the book, as we are in mechanics, then no work has been done on either, since neither has moved.

4.1.4 Varying forces

Now in our definition of work done we have assumed the force exerted to be constant. The force/displacement graph for lifting a brick mass *m* through height *h*, therefore, looks like Figure 25. The work done (*mgh*) is given by the shaded area.

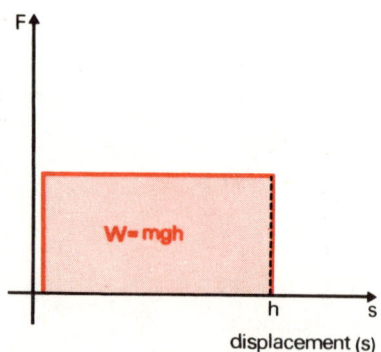

Figure 25 The variation of the force on the brick with displacement. The shaded area is the work done by the force.

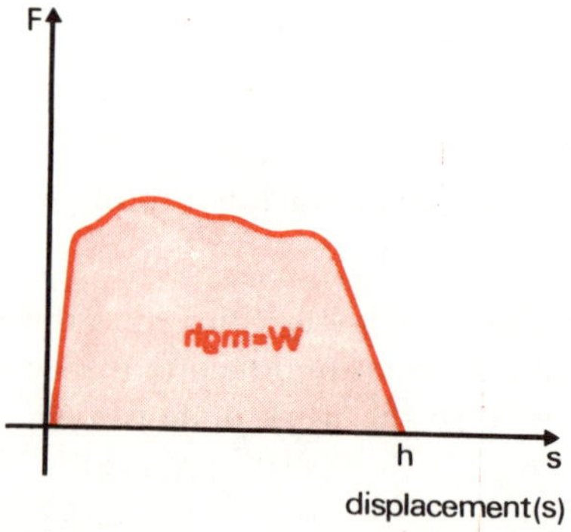

Figure 26 The successive values of a force which varies with displacement. The shaded area still represents the work done by the force.

There is no need for the force involved in 'doing work' to be constant, in general it will not be. In such cases the work done is still given by the area under the curve (Figure 26), the work done being evaluated as an integral, that is

$$W = \int_0^h F \, ds \dots \dots \dots \dots \dots \dots (13)$$

See if you can develop an argument involving a staircase graph to justify this. (You will not need to evaluate an integral of this type in any of the exercises, assignments or self-assessment questions associated with this unit.)

4.2 Something for something

Whenever we do work mechanically, just as in everyday life, we expect to get something from it. So far we have discussed doing work in order to raise objects from one point to another. Frequently we exert a force in order to get something moving. Once an object is moving, as a result of work done, it is capable of doing further work for us: for example, when railway trucks shunt along a line.

Suppose an object of mass m is moving with speed v_0 from right to left (horizontally) but is opposed by a constant force F which acts from left to right. The magnitude of the resulting acceleration we can represent by a. The object will obviously slow down and will lose a m s^{-1} every second. After a time, t, the speed of the object will be v, where

$$v = v_0 - at \dots \dots \dots \dots (14)$$

If we put $v = 0$ in the above equation we obtain the time required for the object to come to rest:

$$t = \frac{v_0}{a}$$

The force $F = ma$ by Newton's second law, so we can substitute for a in the above expression. Then

$$t = \frac{mv_0}{F} \dots \dots \dots \dots (15)$$

The acceleration of the object is uniform so we can work out the distance (d) travelled by the object before coming to rest in terms of its average speed ($v_0/2$) by putting $d = \frac{tv_0}{2}$ and substituting for t from equation (15) giving us

$$d = \frac{mv_0^2}{2F} \dots \dots \dots \dots (16)$$

Equation (16) can be rearranged to give us

$$Fd = \tfrac{1}{2} mv_0^2 \dots \dots \dots \dots (17)$$

Fd is recognizable as the work done by the object in opposing the force F through the distance d (in the same way as the work done by the man in opposing the weight of the brick mg through a height h is mgh). So the expression on the right-hand side of equation (17) is a measure of the capacity of the object, when moving at speed v_0 for doing this amount of work. We say that the *capacity* of an object *for doing work* is its *energy*.

energy

A moving object has *energy* by virtue of its motion. Generally the faster moving or more massive an object, the more energetic (and useful) it is.

The energy an object has by virtue of its motion is called *kinetic energy*. The kinetic energy of a body of mass m moving at speed v is given by half the product of its mass and the square of its speed, symbolically, $\tfrac{1}{2} mv^2$.

kinetic energy

We have concluded that the kinetic energy of a body is a measure of its capacity to do work. Conversely, we would expect that the work done on an object to accelerate it from rest up to a certain speed should equal the kinetic energy of the object at that speed.

Try to prove this yourself.

Hint: assume a constant force F acting on an object of mass m. Find an expression for the speed v and the displacement s after time t. Express the speed in terms of the displacement. Compare the expression for the work done (Fs) with the kinetic energy.

4.3 Storing energy

We started off our discussion of work by considering a man loading bricks on to a lorry. Work has been done on the bricks in this case, but they are stationary on the tailboard of the lorry afterwards. However, if one of the bricks were to fall from the lorry, it would evidently be capable of doing work. The bricks were given this capacity for doing work (energy) when they were raised to the height of the tailboard. We call this type of energy *gravitational potential energy* (potential energy for short). This is energy by virtue of position or height above the surface of the Earth. The potential energy of each brick on the lorry is equal to the work done to get it there (*mgh*).

potential energy

Consider a brick which falls from the lorry. We can follow its subsequent history. It accelerates at g m s^{-2}, so the speed increases by g m s^{-1} every second until it hits the ground. If this occurs after time t, the speed of the brick just before it hits the ground will be gt m s^{-1}. This means that its average speed is $\frac{gt}{2}$ m s^{-1}. The distance through which the brick falls is h metres, so the time t must equal h divided by the average speed

$$t = \frac{h}{\left(\frac{gt}{2}\right)} = \frac{2h}{gt}$$

and $t^2 = \frac{2h}{g}$,

so the final speed $(gt) = \sqrt{2gh}$.

The kinetic energy of the brick at this speed is therefore $\frac{1}{2}m(2gh) = mgh$. In other words, the potential energy lost by the brick equals the kinetic energy gained. We say that the total *energy* (kinetic energy plus potential energy) *is conserved*.

conservation of energy

This is the second important conservation principle that you have come across in working through this unit.

Can you remember the other?

This was the conservation of momentum, which was derived after considering the product of force and time (impulse). The new conservation principle, the conservation of energy, has followed from a consideration of the product of force and distance (work).

Exercise 9

A rock of mass 0·5 kg is thrown vertically into the air with velocity 5 m s^{-1}. What is the mechanical explanation of its subsequent history?

Answer

While moving upwards the rock is slowing down with an acceleration of $-g$. The maximum potential energy the rock can gain is equal to its initial kinetic energy. Let us say that the height it reaches is h m above the point from which it was thrown. At this point the kinetic energy 'runs out'. The conservation of energy then implies that:

$$0·5 \times 9·81 \times h = \tfrac{1}{2} \times 0·5 \times (5)^2$$
$$\therefore h = 1·3 \text{ m (approximately)}$$

At this point the rock has no option but to move downwards again as it still experiences the pull of gravity without anything to oppose it. The kinetic energy the rock can now gain is equal to the potential energy it has at a height of 1·3 m (approximately). Without working this out it should be clear that when it returns to the point of the initial throw its speed is 5 m s^{-1}.

4.4 There's the rub

4.4.1 At the start of Section 4.2, I referred to an object that moved against a constant force and eventually came to rest, working against the force. This is common experience. Whenever we induce motion in an object it eventually comes to rest, irrespective of the kinetic energy we have imparted.

4.4.2 The culprit is *friction*. In the case of movement over a solid surface, when two surfaces rub against each other, the frictional force which opposes the motion is proportional to the force which presses the two surfaces together during the motion. The constant of proportionality (μ) is called the coefficient of friction. So, if we were to drag a box of mass m with force F and the coefficient of friction between the box and the surface was μ, the frictional force would be $-\mu mg$ (the negative sign implies the opposite direction to F.).

friction

Exercise 10

What would be the acceleration of a sliding coin of mass 0·01 kg upon which is exerted a steady force 0·05 N, parallel to a surface with the coefficient of friction between coin and surface of 0·102?

The frictional force in this case has a magnitude $0·01 \times 0·102 \times 9·81 = 0·01$ N. The resultant force on the coin therefore has magnitude $(0·05-0·01)$ N $= 0·04$ N. Using $F = ma$, the acceleration of the coin is $\frac{(0·04)}{0·01}$ m s^{-2} $= 4$ m s^{-2}.

4.4.3 In a fluid the friction which leads to resistance to motion is called *viscosity*. The force opposing motion in a fluid is commonly proportional to the *velocity* of the moving object and as in the case of 'dry' friction, acts in the opposite direction to the motion. In many cases, and particularly at low speeds, the resistive force is directly proportional to velocity ($F = k\mathbf{v}$), where k is some constant of proportionality. At higher speeds, however, and particularly in gases, the force bears a more complicated relationship to the motion because *turbulence* (unsteady movement of the gas), rather than viscosity, plays a major part in the resistance to motion.

viscosity

turbulence

4.4.4 The presence of friction means that we cannot simply equate the work done in moving an object to the kinetic (or potential) energy gained by that object. It also means that the total work done in moving an object is not independent of the path taken, as it was in the case of work done against gravity.

For example, when loading a brick on to a lorry, if the brick is pushed along the back of the lorry, part of the total work done will be that necessary to oppose the force due to friction during the relevant part of the path (Figure 27). The total work done, therefore, will depend upon the path taken.

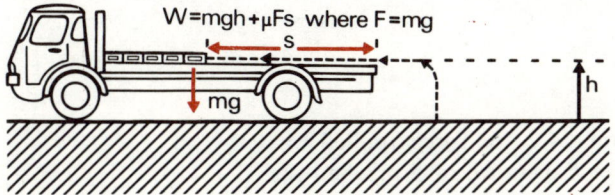

Figure 27 A path taken by the brick which includes work done against friction.

4.5 Simplifying the picture

In many cases it is not at all easy to state how much of the total work done is done against friction, since the frictional force may not be defined. However, where the frictional forces are small and can be ignored, the ideas of work and energy enable us to make straightforward assessments of motion where detailed identification of the other forces involved would be much more difficult.

Exercise 11

A helter-skelter is 10 m tall and a child starts from rest at the top. What is the child's speed at the bottom (ignoring friction)?

> The potential energy (PE) of the child at the top of the helter-skelter is given by $PE = 98 \cdot 1 \, m$ Nm, where m is the mass of the child. This must be equal to the total kinetic energy (KE) gained by the child. So, if the speed of the child at the bottom is v,
> $\frac{1}{2} mv^2 = 98 \cdot 1 \, m$,
> and $v = \sqrt{196 \cdot 2} \text{ m s}^{-1} = 14 \text{ m s}^{-1}$.

(In problems of this type where an object falls, changing potential energy at height h into kinetic energy, the speed at the end of the fall is always given by $\sqrt{2gh}$.)

The ideas of work and energy also enable us to describe the behaviour of a *waterfall*. Suppose we were to fill a bucket with water and then raise it to some height h above the ground. All of the water in the bucket would then be roughly at the same height. If the volume of the bucket was V and the density of the water ρ, the mass of water inside the bucket would be ρV (density is mass per unit volume). If the bucket were to vanish suddenly we would have a mass ρV of water suspended in mid-air, with potential energy $\rho V g h$ resulting from the work done to get it there. This would rapidly be converted into kinetic energy as the water fell, until at the ground the water would be moving at a speed v given by $\sqrt{2gh}$, and have a kinetic energy of $\rho V g h$.

waterfall

Using the waterfall example, we may imagine each 'bucketful' at the top of the fall with potential energy $\rho V g h$ ready to transform this energy into kinetic energy down the fall. This will in fact be a continuous process. If the total volume of water *per second* arriving at the top of the fall is V' then $\rho g V' h$ will be the additional kinetic energy available at the bottom *per second*.

In hydro-electric power stations, the potential energy available from water at a certain height is used to drive turbines. *The rate at which this work is done* is equal to the energy available per second and is called the *power*. In the case of a waterfall it is the rate at which potential energy at the top of the fall is converted into kinetic energy at the bottom.

power

You will find these ideas used on an even larger scale in an estimate of the total energy available on Earth from tidal power. This estimate is on page 210 of the set book, *Resources and Man*.*

4.6 Dissipation, conversion and conservation

4.6.1 In the sliding coin example of Section 4.4.2, the force applied was 0·05 N. If this force were to be applied over a distance of 5 m, then the

---

* US National Academy of Sciences (1969), *Resources and Man*, W. H. Freeman and Co.

work done would be 5×0.05 N m $= 0.25$ N m. However, in this example the acceleration of the coin was 4 m s^{-2}.

What would be the speed of the coin after a distance of 5 m?

> The speed (v) is given by $\sqrt{2as}$ where $a = 4, s = 5$, i.e. the speed is $\sqrt{40}$ m s^{-1}.
> So the kinetic energy $= \tfrac{1}{2} mv^2$
> $\qquad\qquad\qquad\quad = \tfrac{1}{2} \times 0.01 \times 40$ N m
> $\qquad\qquad\qquad\quad = 0.20$ N m (or J).

Comparing this value with the total work done (0·25 N m) it is clear that 0·05 N m of the work done had not been turned into kinetic energy. This must be the work done against friction. This can easily be checked since the frictional force is 0·01 N and over a distance of 5 m the work done against this force must be $5 \times 0.01 = 0.05$ N m. But does this work appear as energy? If so, in what form does the energy appear?

Figure 28 '*Another resulting form of energy is sound*'.

4.6.2 Rub any two objects together and they will both get warm and make a noise. In other words, *heating* is one result of work done against friction and another resulting form of energy is *sound*. When energy is produced in forms which are not useful during an energy conversion process it is said to be *dissipated*. An aim in many technological processes is to reduce *dissipation* to a minimum. However, the part played by sound is often overlooked. The energy content of sound is usually a fairly small fraction of the total energy involved. For example, the energy output from a jet engine in the form of sound amounts to only 1% of the total energy output. From the point of view of dissipation or 'wastage' of energy the noise of a jet engine is unimportant. Nevertheless, in maintaining the environment such noise is an increasingly disturbing factor.

dissipation

4.6.3 To do any kind of mechanical work ourselves we need sustenance. Food supplies chemical energy which we convert into the required form of potential or kinetic energy. If we require a mechanical device to do the tasks, this device in turn needs fuel to generate the necessary energy.

The essential characteristic of mechanical work is that it involves an energy conversion process. You will meet an account of other forms of energy

Figure 29 'This device in turn needs fuel to generate the necessary energy'

and trains of energy transfer in the unit called Energy conversion.

As long as we can name and quantify all the energies involved, then the expressions 'work done' or 'doing work' are redundant. We can replace them by 'energy transferred' or 'transferring energy'. We always have a check on whether or not our quantities of energy are correct through the principle of energy conservation.

4.7 The pile-driver

The two conservation laws which have been presented in the unit can both be used to describe the motion of a pile-driver. Suppose that a pile of mass 500 kg is driven into the ground by a hammer of mass 1 000 kg. Suppose also that, before the blow in which we are interested, the hammer is raised through a height of 2·5 m.

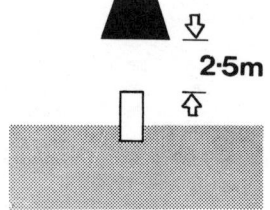

What will be the speed of the hammer before impact?

The potential energy of the hammer before being dropped is $1\,000 \times 9{\cdot}81 \times 2{\cdot}5$ N m. If we represent the speed of the hammer just before impact with the pile by v, then conservation of energy implies that

$$1\,000 \times 9{\cdot}81 \times 2{\cdot}5 = \tfrac{1}{2} \times 1\,000 \times v^2$$

so
$$v = \sqrt{5 \times 9{\cdot}81}$$
$$= 7{\cdot}005 \text{ m s}^{-1}$$

The pile must do work as it is forced through the ground—which naturally resists. So the next question is—what energy is available?

To calculate this energy we must know the speed of the pile after impact. This information follows from the conservation of momentum provided that we make some assumption about the velocity of the hammer after impact. A simple assumption is that the hammer and pile move together.

Before impact only the hammer is moving, with momentum

$$= 1\,000 \times 7{\cdot}005 \text{ kg m s}^{-1}$$

So if the speed of the hammer and pile after impact is represented by v', then

$$1\,000 \times 7{\cdot}005 = 1\,500\, v'$$

and
$$v' = \tfrac{2}{3} \times 7{\cdot}005 \text{ m s}^{-1}.$$

The combined kinetic energy is therefore

$$\tfrac{1}{2} \times 1\,500 \times (\tfrac{2}{3} \times 7\cdot005)^2 = 16\,350 \text{ N m}.$$

4.8 The advanced passenger train

The advanced passenger train consists typically of two power units separated by five carriages. The total mass is approximately 22 000 kg. Information that is of value to the designers is the total power which is necessary to move the train at a certain steady speed up the steepest incline (approximately 1 in 40).

The resistances to motion will include rolling friction at the wheels and air resistance. A fraction of the total power produced will be expended in overcoming these resistances. An estimate of this fraction, at any speed, can be obtained from a graph of the *tractive resistance* experienced by the train against speed, on a level track. Such a graph might be as shown in Figure 30.

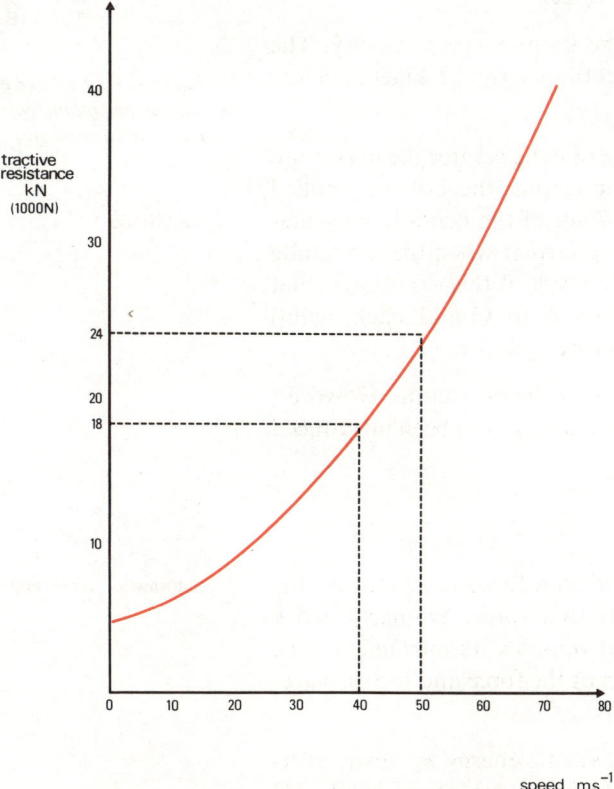

Figure 30 The total resistive force experienced by the advanced passenger train at constant speeds (up to 70 m s^{-1}) on a level track.

Let us suppose that we wish to calculate the total power required to ascend the incline at (a) 40 m s^{-1} and (b) 50 m s^{-1}.

If the force needed to overcome the resistances to motion at speed v is F, then the work done against the resistances every second is Fv. Since power is the rate of doing work, Fv must represent the power needed merely to overcome the resistances. Using the graph, this power is $18\,000 \times 40$ N m s^{-1} at 40 m s^{-1}. The unit of power, N m s^{-1} or J s^{-1}, is called the *watt* (W).

watt

The incline is 1 in 40, this means that for every 40 m of track covered there is a gain in height of 1 m. When the train moves at 40 m s^{-1}, this height is gained each second. The work done is rising by this amount each second, in other words, the power $= 22\,000 \times 9\cdot81 \times 1$ W. Therefore the total power needed $= 935\,820$ W.

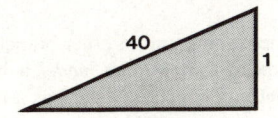

Similarly at 50 m s^{-1}, the power needed to overcome the resistances = 24 000 × 50 W. The power needed to gain a height of $\frac{50}{40}$ = 1·25 m in each second = 22 000 × 9·81 × 1·25 W. The total is 1 469 775 W.

4.9 The pendulum

How would you describe the energy conversion process that takes place during the oscillation of a pendulum?

Consider Figure 31 which represents a pendulum consisting of an inextensible* light string together with a bob of mass m. When the bob is pulled to position A it will be at a height h above its lowest position (B). So, with respect to B it has potential energy mgh. As we know that its total energy must be conserved after release, we can calculate the speed of the bob at B (v). The kinetic energy gained up to B must equal the potential energy 'lost' from A.

$$\tfrac{1}{2} mv^2 = mgh, \quad v = \sqrt{2gh}$$

As the bob is at rest at A and C, v must be its maximum velocity. The energy conversion progression is from potential energy to kinetic energy and back to potential energy again.

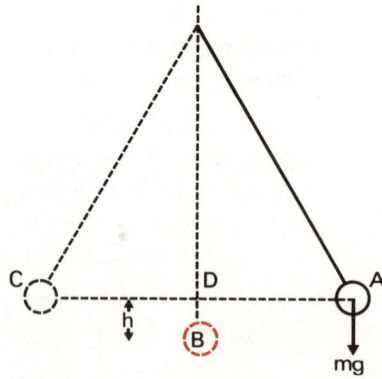

Figure 31 The extremes of swing of a simple pendulum consisting of a bob of mass m on a string.

amplitude

A closer look shows that the greater the value of h the greater the maximum velocity. The height h would be greater the farther the bob was pulled back before release, i.e. the greater the *amplitude* of the pendulum oscillation. So, although the pendulum has to swing farther when the amplitude is increased, the bob moves faster. A possible result of this situation is that the total time for one complete swing (from A to C and back again) does not depend on the amplitude of the swing.

In the next section we will discuss the motion of the pendulum. However, before we do, I will bring together the main ideas that have been introduced concerning work and energy.

Work is done *on* an object (*by* the mover) when a force is exerted on the object and it is moved in the direction of that force. Similarly, work is done *by* a moving object on a force that opposes its movement. The amount of work done is equal to the product of the force and the displacement in the direction of the force.

summary—work and energy

An object (of mass m and with speed v) has kinetic energy by virtue of its motion ($\tfrac{1}{2} mv^2$) and potential energy (mgh) by virtue of its height (h) above a convenient datum level. When work is done, energy is transferred from one form to another or to several other forms.

---

* In mechanics, and in other branches of science, when we call something 'inextensible', we mean that we can *model* it by something which cannot be stretched. Similarly, something which is 'light' has a mass which can be ignored compared with the other masses in the problem.

Section 5

Simple harmonic motion

5.1 Time for a swing

In Figure 31 the distance AD at any time may be referred to as the displacement of the pendulum bob. If you were to plot a displacement/time graph for the bob of the oscillating pendulum starting with the bob in the centre of the swing, you would obtain a graph similar to that in Figure 32 provided the amplitude of oscillation were small. This graph is typical of a *simple harmonic motion*, in which an object moves under a *restoring force* proportional to the displacement of the object from the centre of the oscillation.

Take a close look at the graph and check that its form is consistent with what you already know of the motion of the pendulum. For example, from the energy considerations of Section 4.9, you know that the velocity of the bob is greatest at the centre of the swing.

What can you say about the slope of the displacement/time graph at this point?

Since the slope represents the velocity of the bob, it should have its largest value at this point.

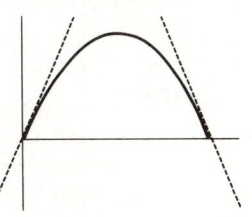

> *Exercise 12*
>
> Draw the forms of velocity/time and acceleration/time graphs corresponding to Figure 32. With these, verify that the accelerating force is proportional to the displacement.

You should find that, although you do not know the actual acceleration at any point, the acceleration/time curve is essentially the mirror image of the displacement/time curve.

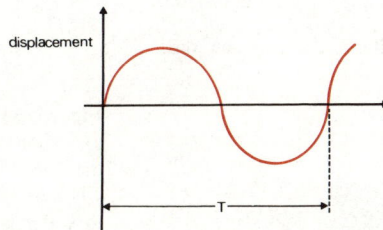

Figure 32 *The displacement of the bob against time, when the amplitude of its swing is small.*

The part of the displacement/time curve marked out in Figure 32 represents one complete oscillation or *cycle*, and the time T marked on the graph is referred to as the *periodic time* or *period* of oscillation. In the simple harmonic motion of a mass m subject to restoring force of magnitude F per unit of displacement, the period T is given by

period

$$T = 2\pi \sqrt{\frac{m}{F}} \dots\dots\dots\dots(18)$$

This means that the period depends only on the acceleration due to the restoring force and is independent of the amplitude, as I implied in Section 4.7. The number of such cycles which take place in one second is referred to as the *frequency* of the motion. For a freely vibrating system like a pendulum this is the natural frequency (Unit 3, Section 2). From the description of T given above, the frequency must equal $1/T$ cycles per second (Hz).

frequency

In the compression (or extension) of a spring the force per unit displacement of the end of the spring is known as the *stiffness* of the spring. From equation (18) it is clear, for example, that for a given mass the frequency of undamped motion will be higher the stiffer the spring.

stiffness

5.2 Damping

If any system is made to oscillate or vibrate it will eventually come to rest. This is similar to our experience of the reduction by friction of the kinetic energy of a body moving in a straight line. In the case of simple harmonic motion with gradually reducing amplitude, the motion is said to be *damped*. The damping of a pendulum is caused by friction at the support and by the resistance of the air through which the bob of the pendulum is moving.

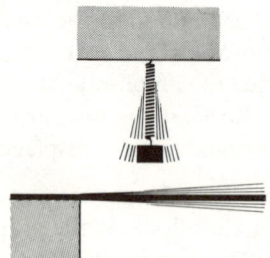

What does a displacement/time curve for a damped oscillation look like?

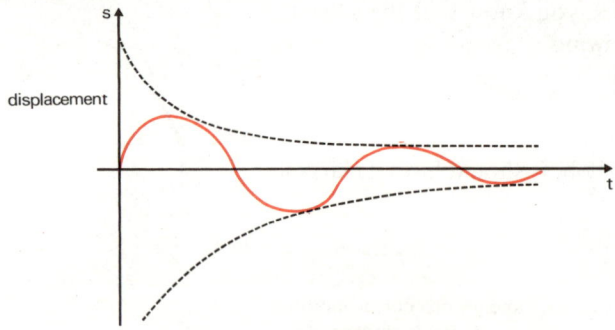

Figure 33   A damped oscillation. The peaks fit on to an exponential curve.

Although the period of the damped vibration of the vibrating system is unaffected by the amplitude of the motion (provided that the amplitude is small), it is different from the period for undamped motion. The typically resulting curve is shown in Figure 33.

> *Exercise 13*
>
> Draw displacement/time curves showing the likely differences between the vibrations of a mass on a spring in air and in water.

Figures 34 and 35. The displacement axis represents vertical displacement measured from the position of the mass when at rest. The decay* of oscillation in water is faster than that in air.

5.3 Strong damping and critical damping

What do you think would happen to the motion if the damping were increased indefinitely?

A point would be reached where even one complete oscillation would be hard to detect. The dotted lines shown in Figure 33 represent *exponential* curves and their steepness determines the rapidity of damping. If you wish to examine the characteristics of the exponential curve you should consult the program GRAPHS in the Supplementary material on computing.

* The word decay is used here in the everyday sense of 'falling-off in quality or quantity'.

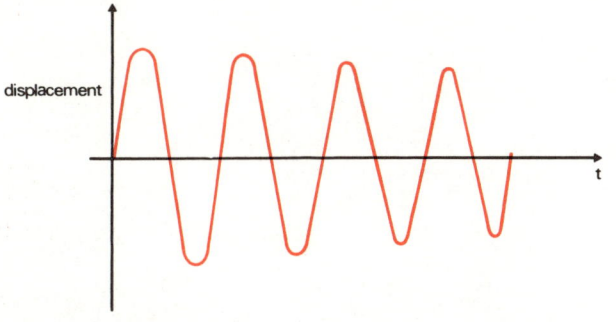

Figure 34 A lightly damped oscillation—for example a mass on a spring vibrating in air.

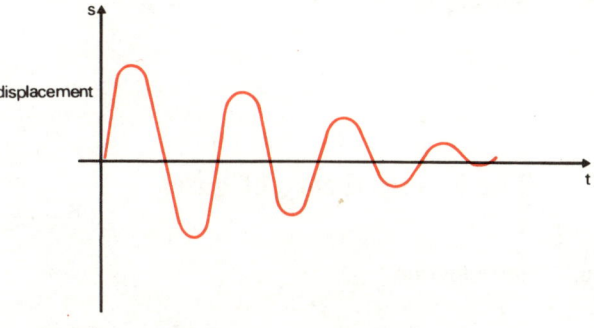

Figure 35 A strongly damped oscillation—for example the mass on a spring vibrating in water.

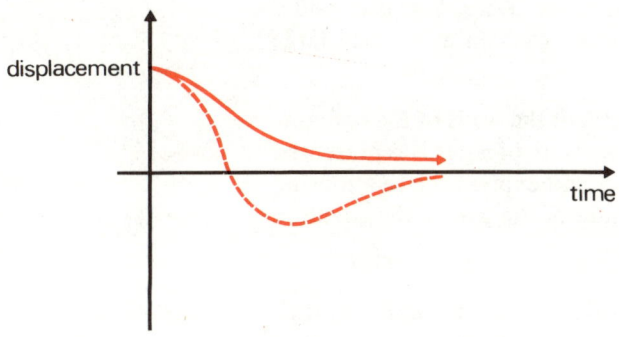

Figure 36 The solid line represents a critically damped vibration. The dotted line represents a vibration which is not quite critically damped. Notice that under critically damped conditions we have to pull the mass off-centre and then release it in order to start any form of motion.

Conditions of strong damping (otherwise referred to as heavy damping) are shown in Figure 36. As the damping is further increased there comes a point at which it is not possible to detect any oscillation. Under this circumstance, if a pendulum, for example, were drawn to one side and released it would return to its original position without swinging further. This is a condition of *critical damping*.

critical damping

If the damping were to be increased beyond the critical point the speed at which the oscillating system would return to its equilibrium position would become less and less. The system would then be *over damped*.

Most machinery contains rotating or reciprocating parts which cause periodic or continuously repeated forces to be applied to the structure of the machine. These forces cause the machine to oscillate or vibrate. Excessive vibrations may damage the support on which the machine rests, may cause undue wear in the moving parts (T100/BK 1 p. 45) or may be responsible for the emission of noise.

Periodic motion occurs widely in nature, ranging over such examples as the daily rotation of the Earth, the annual changes in the seasons, the tides, the vibrations of an insect's wing and the heart beat. The engineer's attitude to vibrations is determined by the purpose for which the mechanism under consideration is designed. In a clock the vibrations of a pendulum or a balance wheel must be maintained in order to produce an oscillation of fixed period; the energy put into the system must be just sufficient to overcome the energy dissipated through friction (damping). On the other hand a car spring must protect the car and its occupants from sudden jolts. However, the initial oscillations set up must be damped as quickly as possible by shock absorbers, which dissipate the energy of the spring.

Section 6

The kinetic theory of gases

6.1 Introduction

If a closed container containing a gas is heated, the enclosed gas expands, exerting pressure on the sides of the container. If heating continues the pressure will continue to rise until the container bursts—this is why it is so dangerous to throw aerosol containers on to a fire. Let us consider the mechanics by means of which the enclosed gas can exert such large forces.

We call the force exerted by a gas per unit area of the walls of the containing structure, the *pressure* of the gas. If the pressure of a gas is represented by the letter P and its volume by V, then we can express the relationship between the pressure, volume and temperature of the gas by the equation

$$PV = rT \quad \ldots\ldots\ldots\ldots\ldots\ldots(19)$$

pressure

T is the *temperature* expressed on the scale called the *absolute scale* (Figure 37). You will have met this relationship and some of the reasoning behind it in Modelling I. It is called the *ideal gas law*, and the kinetic theory of gases sets out to develop models that explain these observations. It is interesting both as an example of modelling and because it brings together several of the ideas that we have met and defined already in this unit.

temperature

ideal gas law

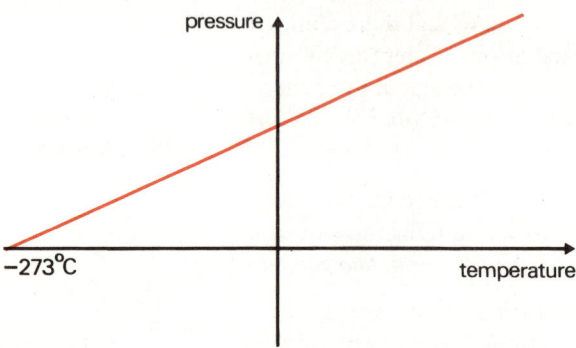

Figure 37 The variation of pressure with temperature for an ideal gas.

In one of the simplest models of the theory, a gas is taken to consist of small identical rapidly moving particles (*molecules*) which collide with the walls of the container but not against each other.

molecules

In the next sections I present a version of this model and an associated theory that are similar to ones originally proposed by a physicist called Joule.

6.2 Keeping up the pressure

Each particle, when it collides with the walls of the container, will exert a small impulse. There are a great many of these particles or molecules in the gas and so they will make contact with the walls of the container in rapid succession. Each impulse has a characteristic force/time curve similar to that shown in Figure 38. If you plotted the force/time curve for a large number of these impulses occurring in rapid succession it

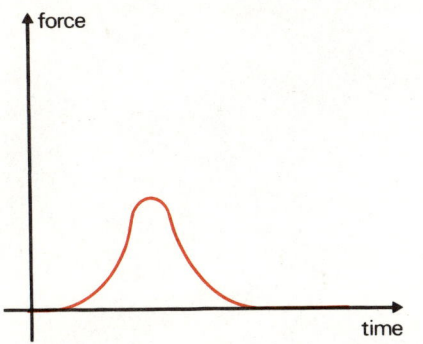

Figure 38 *The form of the impulse resulting from the collision of a molecule with a container wall.*

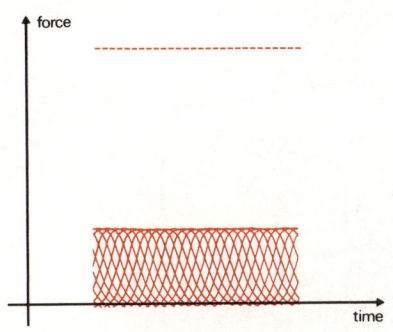

Figure 39 *The addition of a large number of rapidly occurring impulses to produce a steady force (represented by the dotted line).*

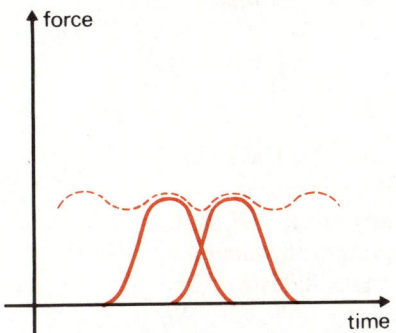

Figure 40 *The addition of two successive impulses. The dotted line represents the variation of the composite force with time.*

might well look like Figure 39. Each impulse curve would be barely distinguishable from the next. To see how these impulses would add together, let us look at an example where only two impulses are concerned, shown in Figure 40. As the two impulses overlap (part of the time they are both acting at once) it is only necessary to add the ordinates together during the period of overlap in order to get the total force exerted shown by the broken line in Figure 40. The result of adding the ordinates of a large number of overlapping impulses would be an almost steady force as represented by the broken line in Figure 39.

Figure 41 '*A large number of impacts can produce a steady force.*'

6.3 From change in momentum to pressure

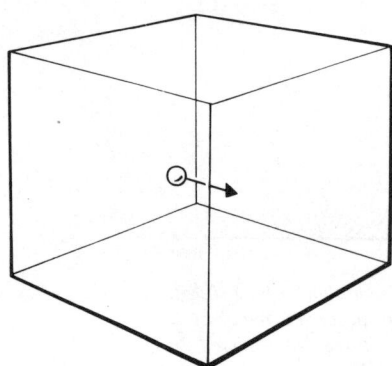

Figure 42 A molecule moving towards the right-hand wall of a cubic container.

Suppose that the container is a cube-shaped box (Figure 42). The simplified model is based on the assumption that, at any given moment, one-sixth of the molecules (all of mass m) are moving towards any given wall of the cube. Let us look at a particular molecule which is moving with velocity v_1 towards the right-hand wall of the cube. If this molecule collides with the wall, we want to know the change in momentum of the molecule. We assume that as a result of the impact, the speed of the molecule (v_1) is unchanged but the direction of its velocity is reversed. We can now do a quick sum to find what change in momentum this represents:

$$\begin{aligned}\text{momentum before impact} &= mv_1 \\ \text{momentum after impact} &= -mv_1 \\ \text{change in momentum} &= mv_1 - (-mv_1) = 2mv_1\end{aligned}$$

Suppose that there are n_1 of these molecules with velocity v_1. The farthest that any of these molecules could be away from the right-hand wall is the width of the cube. If this width is L, any of these molecules can cross the cube in time $\frac{L}{v_1}$. So within the period $\frac{L}{v_1}$ the total change in momentum is $2n_1 mv_1$. This represents the change in momentum of n_1 molecules which collide with the right-hand wall of the cube.

The rate of change of momentum of these molecules is

$$2n_1 mv_1 \div \frac{L}{v_1} = \frac{2n_1 mv_1^2}{L}$$

(This is the change in momentum of n_1 molecules divided by the period during which it occurs.) You should remember that the rate of change of momentum is one of our definitions of force from Newton's second law. So $\frac{2n_1 mv_1^2}{L}$ represents the force on the right-hand wall of the cube because of the impacts of the n_1 molecules with speed v_1. Suppose that in addition to these molecules there are n_2 with velocity v_2, and n_3 with velocity v_3 and so on. We could use exactly the same argument to show that the force exerted by n_2 molecules with speed v_2 is $\frac{2n_2 mv_2^2}{L}$, the force exerted by n_3 molecules with speed v_3 is $\frac{2n_3 mv_3^2}{L}$ and so on.

In order to get the total force per unit area, which is the pressure on the right-hand wall, we have to divide each of these expressions by L^2, the area of one face, and add the results.

So

$$P = \frac{2n_1 m v_1^2}{L^3} + \frac{2n_2 m v_2^2}{L^3} + \frac{2n_2 m v_3^2}{L^3} \cdots$$

$$= \frac{2m}{L^3}(n_1 v_1^2 + n_2 v_2^2 + n_3 v_3^2 \cdots) \quad \ldots \ldots (20)$$

The expression in the brackets is extremely cumbersome. However, we know that if the total number of molecules within the box is N, the total number moving towards the right-hand wall of the box is $\frac{N}{6}$.
This means $(n_1 + n_2 + n_3 + \ldots) = \frac{N}{6} \quad \ldots \ldots \ldots (21)$

To make this averaging process clear let us consider first an algebraic expression for *mean speed*. To find the mean of any array, we add together the individual elements and then divide by the total number of elements. With n_1, molecules moving at speed v_1, n_2 at speed v_2 and so on, then the mean speed

mean speed

$$\bar{c} = \frac{n_1 v_1 + n_2 v_2 + n_3 v_3 + \ldots}{n_1 + n_2 + n_3 + \ldots} \quad \ldots \ldots (22)$$

In the same way we can find the *mean square speed*

mean square speed

$$\overline{c^2} = \frac{n_1 v_1^2 + n_2 v_2^2 + n_3 v_3^2 + \ldots}{n_1 + n_2 + n_3 + \ldots} \quad \ldots \ldots (23)$$

Now, as equation (21) tells us $(n_1 + n_2 + n_3 + \ldots) = \frac{N}{6}$

so

$$\overline{c^2} = \frac{n_1 v_1^2 + n_2 v_2^2 + n_3 v_3^2 + \ldots}{\frac{N}{6}}$$

or

$$(n_1 v_1^2 + n_2 v_2^2 + n_3 v_3^2 + \ldots) = \frac{N}{6}\overline{c^2} \quad \ldots \ldots (24)$$

After that slight diversion, we can return to the expression for the pressure on the right-hand wall of the cubic container (equation (20)) and using equation (24) we have

$$P = \frac{2m}{L^3} \frac{N}{6} \overline{c^2} = \frac{Nm\overline{c^2}}{3L^3} \quad \ldots \ldots (25)$$

Not surprisingly $c_{rms} = \sqrt{\overline{c^2}}$ is called the *root mean square speed* (rms speed for short). With so many molecules it is reasonable to assume that the mean square velocity of the molecules is the same whichever wall of the cube they are moving towards. This is tantamount to saying that the pressure on any of the walls is the same.

root mean square speed

Other less restrictive analyses can be carried out but most of them are similar to the one you have just read, and they give the same answer. For example, the assumption that any molecule moves directly towards a particular wall is not a good one. The velocity of any molecule could be in any direction. However, this velocity could be split up into components directed towards the walls of the cube. Three components would be necessary in this case as there are three dimensions in which the molecules can move. The subsequent argument would then be similar to the one above.

6.4 Meaning of temperature

L^3 is the volume of the cube (V). For a container of any shape and volume V we can write the equation

$$PV = \tfrac{1}{3} Nm \overline{c^2} \quad \ldots\ldots\ldots\ldots (26)$$

So the model has shown directly that, for a certain number of molecules with a certain mean square speed, pressure is inversely proportional to volume. Mind you, this could have been argued qualitatively from the model without mathematical analysis (as follows). The pressure on the walls of a container will depend upon the number of molecular impulses per unit time. The number of impulses will be dependent upon the dimensions of the container; the less distance the molecules have to cover between impacts, the more frequent will be their impacts with the walls. This implies that the pressure upon the walls of the container will increase as the volume of the container decreases ($P \propto \dfrac{1}{V}$).

So, if the model can tell us all this qualitatively, why have we bothered to set up a mathematical model? Well let us take a closer look at the right-hand side of equation (26)

$$\tfrac{1}{3} Nm \overline{c^2} = \tfrac{2}{3} N \frac{(m\overline{c^2})}{2}$$

by a little algebraic sleight-of-hand. This expression is two-thirds of the *total kinetic energy* of all the molecules and the part of the expression in brackets is the *mean kinetic energy* of the molecules. (You should check this from the definition of mean square velocity.) So if the total kinetic energy of the molecules inside a constant volume is increased then the pressure is increased. But we also know from empirical evidence that the pressure inside a constant volume enclosure increases as the temperature increases. So the mathematical model suggests a relationship between *mean* kinetic energy (since N is constant if gas is not being added or subtracted) and temperature.

total kinetic energy
mean kinetic energy

Our model accounts not only for the gas law but also gives us some insight into the meaning of temperature. It is qualitatively consistent, since if temperature is proportional to mean kinetic energy, then as temperature increases the mean kinetic energy must increase. This means that the molecules would move around more rapidly and collide with the walls of the container more frequently. By an earlier argument that means an increase in pressure.

Exercise 14

Suppose we have two similar containers of two different gases at the same temperature. What can the kinetic theory tell us about the mean kinetic energy of the molecules in the containers?

If the temperatures of the two gases are the same, then the mean kinetic energy of the gas molecules must be the same. In other words, if the mass of one molecule of the first gas is m_1 and the mass of one molecule of the other gas is m_2 and their mean square speeds are $\overline{c_1^2}$ and $\overline{c_2^2}$ respectively then
$$\tfrac{1}{2} m_1 \overline{c_1^2} = \tfrac{1}{2} m_2 \overline{c_2^2}$$

What do you think would happen, according to this model, if two containers containing gases at different temperatures were joined into one?

Both experiment and intuition suggest to us that the temperatures move towards one another until the temperature in the container is uniform. There is nothing in the previous argument to suggest an explanation. The model we have used so far is inadequate because of one of our initial assumptions, that which suggests that the molecules do not collide with each other. It has to be revoked without losing the idea that temperature is a measure of the mean kinetic energy of the molecules. For it is fairly obvious that the molecules of the gas at the greater temperature will have the greater mean kinetic energy and if they *do* collide with the molecules of the other gas they will quickly transfer some of their energy. Eventually the transfer of the energy will be such that all the molecules will have the mean kinetic energy corresponding to the final temperature of the mixture. You will see the value of this insight into the nature of temperature when you come to the units on chemistry and energy conversion.

Exercise 15

Imagine two different containers of the same size, one of which contains a gas at the pressure P_1 and the other contains a second gas at the pressure P_2. What would happen if the gas in the second container were forced into the first?

Clearly, when a single vessel contains both types of molecules, the pressure on the walls of the container will be due to the impact of both types of molecules and the pressures will be added together. The total pressure will be $P_1 + P_2$. Do you think that this can be extended to a mixture of three or more gases?

6.5 A few complications

It was quite unnecessary to confine the consideration to a cubic container. Experiment shows that Boyle's law holds for gas in a container of any shape or size, and our argument could have been extended to take account of this, but the analysis would not have told us anything new.

We have already considered the assumptions that all the molecules within the gas of the containing vessel are identical (Exercise 15) and also that the molecules do not collide with each other. However, the calculation concerning momentum change requires the molecules to behave like small elastic spheres. In addition, we have assumed the molecules to have only kinetic energy by virtue of motion in a line. We have not allowed them to spin or vibrate. All of these assumptions are only strictly true if we are considering a *monatomic gas* such as neon.

monatomic gas

In addition to colliding with each other, the molecules have an appreciable effect on each other's motion since they occupy a finite volume and exert mutual forces called *van der Waal's forces*. There is a further discussion of van der Waal's forces in the unit on Atoms and molecules.

van der Waal's forces

Self-Assessment Questions

Section 3.1

SAQ 1 (Objectives 1, 2)

In a head-on crash, hinged bucket seats:

(a) tend to tip their occupants into the windscreen;

(b) try to rotate backwards through the car floor;

(c) are no more dangerous than fixed seats.

Which of these statements is most correct?

Section 3.2

SAQ 2 (Objectives 2, 8)

Which of Newton's laws of motion accounts for the mechanism of rocket propulsion?

SAQ 3 (Objectives 1, 2)

What is the tension in the lift-cable when a lift of mass 1 000 kg is accelerating (steadily) upwards at 2 m s^{-2}?

Section 3.3

SAQ 4 (Objectives 1, 2, 8)

If you cue a billiard ball against a cushion at right-angles, and stop it with your hand after the rebound

(a) how many impulses does the ball undergo?

(b) which is the greatest impulse?

SAQ 5 (Objectives 1, 2, 8)

What should be the object of a crash barrier on a motorway?

(a) to bounce an uncontrollable vehicle back into the traffic stream?

(b) to reduce the speed of the vehicle so that it crosses the central reservation at a reduced speed?

(c) to stop the vehicle altogether on the central reservation?

Compare these alternatives and discuss the design relevant to your choice in terms of force and momentum.

SAQ 6

Which of the following statements is true of a collision between two objects?

(a) The total momentum before and after the collision is always the same.

(b) The total momentum before and after the collision is never the same.

(c) The total kinetic energy before and after the collision is always the same.

(d) The rate of change of momentum is conserved.

Section 3.6

SAQ 7 (Objectives 1, 7)

There are two satellites of the same mass which are moving around the Earth in near circular but different orbits.

Which of the following statements are true?

(a) The satellite in the higher orbit moves faster than the other satellite.

(b) The satellite in the higher orbit moves more slowly than the other satellite.

(c) The centripetal force needed to keep the satellite in the higher orbit is larger than that for the other.

(d) The centripetal forces must be the same for both satellites.

(e) The centripetal force needed to keep the satellite in the higher orbit is smaller than that for the other.

SAQ 8 (Objectives 1, 2, 4)

An early instrument of war was the Greek fighting galley (Figure 43). The name 'galley' is applied to all those warships which were propelled chiefly by oars. A particular class of fighting galley, and the fastest, was the trireme. There were no missile weapons effective enough to sink a ship and so the only practicable form of attack was to ram. The ram projected several feet ahead of the ship's stem and was heavily sheathed in cast bronze. Given that the total *weight* of the ship was approximately 40 000 N and that its maximum speed was 10 m s^{-1}, calculate the kinetic energy of the trireme at full speed.

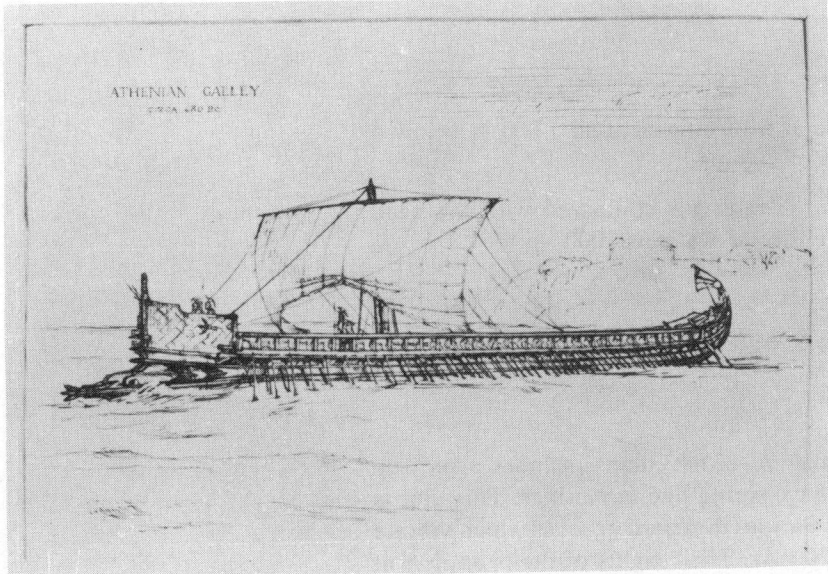

Figure 43 A picture of a Greek fighting galley.

Section 4.4

SAQ 9 (Objectives 1, 9)

In a car, power is produced in the engine by consumption of petrol. The movement of the pistons within the engine is transmitted to the wheels through the clutch, the gear box and the rear axle, and from the wheels to the road.

(a) Where should the frictional forces be made as high as possible?
(b) Where should the frictional forces be minimized?

SAQ 10 (Objectives 1, 3, 4, 9)

A railway wagon of mass 50 000 kg is at rest on a level stretch of track. If the resistances to motion are equivalent to a coefficient of friction of 0·005, determine:

(a) the force which would be required to set the wagon rolling with a steady acceleration of 1 m s^{-2};

(b) the work done by this force in moving the wagon 150 m;

(c) the gain in kinetic energy of the wagon;

(d) the difference between (b) and (c), and explain what this difference represents.

SAQ 11 (Objectives 1, 7, 9)

What is the maximum speed at which a car can safely negotiate a bend of radius 75 m, given that the coefficient of friction between the road surface and the car tyres is 0·9. Assume that the road surface is level, that the car skids before it tilts or overturns and that the frictional force acts radially. (You will find an additional self-assessment question associated with Section 4.4 if you consult the Supplementary material on computing.)

Section 4.5

SAQ 12 (Objectives 1, 5)

What is the total power available at a turbine situated 1 000 m below a reservoir which empties at a rate of $200 \text{ m}^3 \text{ s}^{-1}$?

You can disregard the depth of the reservoir compared with the head from reservoir to turbine. The density of water is 1000 kg m^{-3}.

Section 5

SAQ 13 (Objective 10)

The transcription unit in a number of hi-fi systems includes a pick-up arm and a stylus, both of which have spring-like mountings. This unit is designed to be able to follow irregularities in record grooves which vibrate the pick-up at frequencies up to 20 kHz. The weights of the pick-up arm and stylus and the stiffness of their mountings must be chosen so that their natural frequencies do not fall within the audio-frequency range (that is the frequency range which we can hear—see Unit 2, The human component), otherwise the sound generated would be distorted.

The mass of the pick-up arm is, of necessity, much greater than that of the stylus.

How would you choose the stiffnesses of their mountings?

Section 6

SAQ 14 (Objectives 1, 11)

Why is it wrong to check your car tyre pressures when the tyres are warm?

Section 6.3

SAQ 15 (Objective 11)

Three helium molecules are moving with speeds of 400, 500 and 600 m s^{-1} respectively.

(a) Show that their root mean square speed differs from their average speed.

(b) What is their mean kinetic energy?

(The mass of a helium molecule is $6\cdot 64 \times 10^{-27}$ kg.)

Self-Assessment Answers and Comments

SAQ 1

Statement (c) concurs most with the laws of motion. It would be true to say that a bucket seat will move during an impact in relation to the rest of the car, whereas a fixed seat would not necessarily move. However, the bucket seat will have exactly the same forward velocity as its occupant just before the impact. During the impact, as the seat experiences the same deceleration as its occupant, it will again have the same forward velocity. Therefore the seat cannot produce any *extra* (forward) force. (However, if you are using seat-belts bolted to the car frame, the hinged seat, since it moves forward with you, would exert an additional force on the belt.)

SAQ 2

Newton's third law indicates the mechanism. The analysis of rocket motion can be extremely complicated when such factors as gravitational effects, rocket orientation and the construction of the rocket are taken into account. A simple model would be that of a cylinder closed at one end in which fuel is burnt and ejected through the open end. The ejection of the gases resulting from combustion of the fuel constitutes an action. Therefore, according to Newton's third law, the rocket must experience an equal and opposite reaction—which acts as the driving force. The greater and faster the flow of gas from the rocket, the greater the force.

SAQ 3

The weight of the lift (mg) is $1\,000 \times 9 \cdot 81$ N $= 9\,810$ N. The force needed to accelerate the mass of the lift at 2 m s^{-2} is

$$1\,000 \times 2 = 2\,000 \text{ N}$$

The total force is composed of the force needed to produce the acceleration and the force needed to counteract the weight. So the total force is

$$(2\,000 + 9\,810) \text{ N} = 11\,810 \text{ N}$$

SAQ 4

(a) The billiard ball undergoes three impulses: (1) when cued, (2) on impact with the cushion, (3) on being stopped by your hand. In impulse (1) the speed of the ball (of mass m) is increased from zero to some value v. The momentum change is mv. In impulse (2) we can assume that the ball comes off the cushion so the direction of its velocity is reversed. The change in momentum towards the cushion is

$$mv - (-mv') = mv + mv'$$

where v' is the new speed of the billiard ball. In impulse (3) the new momentum (mv') is destroyed, so the change in momentum is mv'.

(b) Since the impulses are equal to the changes in momentum, the second impulse—the impulse with the cushion—must be the greatest.

SAQ 5

Clearly the safest choice is to stop the vehicle on the central reservation. Other than this choice, the first alternative would be preferred to the second since collision between vehicles travelling in the same direction means smaller momentum change and therefore smaller forces in a certain time interval than collision between vehicles travelling in opposite directions.

To stop a heavy vehicle travelling at high speed means the destruction of considerable momentum. To try and reduce the forces involved, the collision with the crash barrier must be a cushioned one. One way this may be achieved is to support the crash barriers on posts so constructed and spaced that they snap off in succession as the force which they experience exceeds a certain value. In this way, the total impulse is broken up into a number of stages over a fairly long time-interval. This system is prohibitively expensive as it means that the crash barrier can only be used once. A system designed according to (a) is more usual as it requires only minor deformability on the part of the barrier.

Figure 44 A crash barrier designed to prevent any vehicle from crossing the central reservation.

Refer also to: 'Barrier cables stop cars safely' *New Scientist* 6 January 1972

SAQ 6

Statement (a) is correct. In a collision the total momentum is conserved.

SAQ 7

Statements (b) and (e) are correct.
Newton's law of universal gravitation states that the gravitational force between the Earth and the satellite in an orbit of radius R is $\frac{Gmm_E}{R^2}$
This must equal the centripetal force $\frac{mv^2}{R}$, where v is the speed of the satellite. So $\frac{Gmm_E}{R^2} = \frac{mv^2}{R}$ or $v^2 = \frac{Gm_E}{R}$. As R increases, therefore $\frac{Gm_E}{R}$ must decrease—in other words v must decrease.
This means that the satellite in the higher orbit moves at a slower speed.

Furthermore as v decreases and R increases, the centripetal force $\frac{mv^2}{R}$ must decrease. Therefore the satellite in the higher orbit requires a smaller centripetal force.

SAQ 8

Kinetic energy is given by half the product of the mass and the velocity squared. Hence the kinetic energy of the trireme $= \frac{1}{2} \times \frac{40\,000}{9\cdot 81} \times (10)^2$

This is approximately equal to

$$200\,000 \text{ J}$$

which is equivalent to about 5 kg of gunpowder. The impact of the trireme's ram was very effective.

SAQ 9

(a) The clutch plates and the wheels.

As the clutch plates transmit the power from the engine there must be high frictional forces between them. A slipping clutch results in loss in the transmission of power.

High frictional forces are also necessary both to maintain good tyre adhesion with the road and for braking purposes.

(b) The engine, gear box and rear axle.

These are points at which frictional forces must be minimized in order to reduce power losses.

SAQ 10

(a) The resistance to motion must be equivalent to a frictional force of $0\cdot005 \times 50\,000 \times 9\cdot81 = 2\,452$ N. This is the magnitude of the force required just to overcome the resistance to motion. In order to accelerate the wagon at 1 m s^{-2} an additional force of $50\,000 \times 1 = 50\,000$ N is required. The total force required is therefore 52 452 N.

(b) The work done by this force through a distance of 150 m is $52\,452 \times 150 = 7\,867\,800$ Nm.

(c) The speed of the body after starting from rest and moving a distance of 150 m at an acceleration of 1 m s^{-2} = $\sqrt{2 \times 1 \times 150} = 10\sqrt{3}$ m s^{-1}. The kinetic energy of the wagon at this speed is $\frac{1}{2} \times 50\,000 \times 300 = 7\,500\,000$ Nm.

(d) The difference between (b) and (c) is 367 800 Nm which must represent the work done in overcoming the resistance to motion.

SAQ 11

The centripetal force that is required to maintain the car's movement around the bend is provided by the frictional force between the car tyres and the road.

$$\text{So, as } mg = \frac{mv^2}{r}, \quad v = \sqrt{gr} = \sqrt{0\cdot 9 \times 0\cdot 81 \times 75}$$

$$v = 25\cdot 7 \text{ m s}^{-1}$$

SAQ 12

The mass of 1 m³ of water is equal to the density multiplied by the volume. The potential energy of 1 m³ of water in the reservoir is

$$1 \times 9 \cdot 81 \times 1\,000 \text{ J}$$
$$= 9\,810 \text{ kJ}$$

This is the energy available from each 1 m³ of water at the turbine. However, there will be 200 m³ arriving at the turbine each second. So the total energy available per second—or power

$$= 200 \times 9\,810\,000 \text{ W}$$
$$= 1\,962\,000 \text{ kW} \quad \text{or} \quad 1\,962 \text{ MW}$$

SAQ 13

Usually the resonant frequency of the pick-up arm is chosen to lie at as low a frequency as possible (8–25 Hz). The pick-up arm is fairly massive, but the weight of the arm must be limited to the suggested playing weight of the unit. The equation $T = 2\pi \sqrt{\dfrac{m}{F}}$ suggests therefore that the stiffness of the pick-up mounting should be as low as possible (the period is the inverse of the frequency). On the other hand the stiffness of the stylus mounting should be high in order to ensure a small period (high natural frequency).

SAQ 14

When the car is moving along the road, the continual flexing of the tyre walls creates hysteresis effects on the rubber which consequently becomes warmer. As the tyre increases in temperature, the air inside also becomes warmer and therefore the pressure increases until an equilibrium value is reached.

Manufacturers' recommended tyre pressures are based upon 'cold' values.

SAQ 15

(a) The root mean square speed (c_{rms}) is $\sqrt{\dfrac{(400)^2 + (500)^2 + (600)^2}{3}}$

$$= \sqrt{\dfrac{77}{3}} \times 10^2 = 509 \text{ m s}^{-1}$$

The mean speed $= \dfrac{400 + 500 + 600}{3} = 500 \text{ m s}^{-1}$.

(b) The mean kinetic energy of the molecules is $\tfrac{1}{2} m \overline{c^2}$.
So if $m = 6 \cdot 64 \times 10^{-27}$ kg

$$\tfrac{1}{2} m \overline{c^2} = \tfrac{1}{2} \times 6 \cdot 64 \times 10^{-27} \times \dfrac{77}{3} \times 10^4 = 8 \cdot 525 \times 10^{-22} \text{ Nm}.$$

T100 THE MAN-MADE WORLD
Technology Foundation Course Units

Week number	Correspondence text	Unit number
1	Systems	1
2	The human component	2
3	Speech, communication and coding	3
4	Statistics (first part of)	10
5, 6	Production systems modelling / The production environment	28, 29
7	Systems File	5
8	Mechanics	6
9, 10	Electricity and magnetism	7, 8
11, 12	Energy conversion / Power and society	20, 21
13	Environment File	23
14, 15	Maintaining the environment	26, 27
16	Noise abatement	26/27S
17	Economics File	11
18	Structures and microstructures	9
19	Materials	22
20	Chemical reactions	24
21	Chemical processes in industry	25
22	Cities File	30
23	Automatic computing	12
24	The heart of computers	13
25	Computer systems	14
26	Economics of traffic congestion	31
27	Transport File	31 File
28	Reliability (second part of)	10
29, 30	Analogue computing	15
31	Control	16
32	Modelling II	18
33, 34	Design	32, 33, 34

Acknowledgements

Acknowledgement is made to the following sources for illustrations used in this unit:

Figure 4: Engineering Laboratory Equipment Ltd.; Figure 20: Paul Popper Ltd.; Figure 43: Science Museum; Figure 44: Road Research Laboratory.

Further reading

Jackson, A. (1971), *Mechanical Engineering Science for 01*, Longman.

Harrison, R. D. (1972), *Forces*, Longman Physics Topics (revised edition).

Jardine, J. T. (1969), *Mass in Motion*, Longman Physics Topics.